The Dr. James Fair Heritage Distillation Symposium

Topical Conference at the 2011 AIChE Spring Meeting and 7th Global Congress on Process Safety

Chicago, Illinois, USA
13-17 March 2011

ISBN: 978-1-61782-896-6

Printed from e-media with permission by:

Curran Associates, Inc.
57 Morehouse Lane
Red Hook, NY 12571

Printed by Curran Associates, Inc. (2011)

AIChE
3 Park Avenue
New York, NY 10016-5991

Phone: (203) 702-7660
Fax: (203) 775-5177

www.aiche.org

Additional copies of this publication are available from:

Curran Associates, Inc.
57 Morehouse Lane
Red Hook, NY 12571 USA
Phone: 845-758-0400
Fax: 845-758-2634
Email: curran@proceedings.com
Web: www.proceedings.com

TABLE OF CONTENTS

(9a) Chemical Engineering Contributions of Jim Fair

A. Frank Seibert, Separations Research Program, The University of Texas at Austin, Austin, TX and R. Bruce Eldridge, Chemical Engineering, The University of Texas at Austin, Austin, TX

In October 2010, the chemical engineering community lost a great friend and role model with the passing of James R. Fair. The timeless accomplishments of James R. Fair in the areas of distillation, absorption and stripping are well-known within the AIChE distillation community. However, Jim was also equally interested in other chemical engineering areas and provided significant contributions in heat transfer, liquid and supercritical extraction, reactive distillation and adsorption. Jim Fair's accomplishments in these and some surprising areas will be presented.

See more of this Session: Following Dr. Fair's Footsteps: Fractionation and Tray Optimization
See more of this Group/Topical: Topical 8: The Dr. James Fair Heritage Distillation Symposium

(9b) Novel Configurations Using Fewer Columns for Multicomponent Distillation

Anirudh A. Shenvi, Vishesh H. Shah and Rakesh Agrawal, School of Chemical Engineering, Purdue University, West Lafayette, IN

The synthesis of optimal distillation sequences for a given application is a challenging problem in process design. Several researchers have provided solutions for the problem of synthesizing the complete search space of distillation configurations that use exactly (n-1) distillation columns to separate a zeotropic feed mixture into n product streams. However, the search space of all possible distillation configurations that use less than (n-1) distillation columns for an n-component separation has not been completely explored. We have developed a mathematical framework to synthesize the complete set of distillation configurations that use less than (n-1) columns for an n-component separation.

While several configurations with less than (n-1) columns cannot make arbitrarily high purity product streams, there exist a few distillation configurations that use less than (n-1) columns while still retaining the ability to make products with any pre-specified purity. Such fewer column configurations deserve special attention and are derived from the set of configurations that use exactly (n-1) distillation columns by simultaneous heat and mass integration between two or more distillation columns. The Brugma and Kaibel configurations are probably the most well known among such configurations. We present a systematic procedure to synthesize the complete set of fewer column distillation configurations obtained by simultaneous heat and mass integration of columns from the set of configurations with exactly (n-1) distillation columns.

Furthermore, we have found that these heat and mass integrated configurations lead to an entire class of hitherto unknown distillation configurations. These novel configurations show great promise for reducing the energy consumption. A synthesis procedure for these novel configurations will also be described.

See more of this Session: Following Dr. Fair's Footsteps: Fractionation and Tray Optimization
See more of this Group/Topical: Topical 8: The Dr. James Fair Heritage Distillation Symposium

(9c) Minimum Reflux Calculation for Non-Sharp Splits In Multicomponent Distillation Systems Using Column Profile Maps

D. Hildebrandt[1], Ronald Abbas[2] and Chan Yee Ma[2], (1)Chemical and Metallurgical Engineering, Wits University, Johannesburg, South Africa, (2)School of Chemical and Metallurgical Engineering, Centre of Material and Process Synthesis (COMPS), Johannesburg, South Africa

Minimum Reflux Calculation for Non-sharp Splits in Multicomponent Distillation Systems using Column Profile Maps.

Ma, Chan Yee; Abbas, Ronald*; Hildebrandt, Diane

Centre of Material and Process Synthesis

School of Chemical and Metallurgical Engineering

University of the Witwatersrand Johannesburg

Private Bag 3

WITS 2050

Although distillation is a highly energy intensive process, it is the most widely used in industrial separation techniques. Thus, minimizing the energy consumption in a distillation column is of interest, both environmentally and economically. In a distillation process, the minimum energy corresponds to the minimum reflux, or boil-up ratio. Previous minimum reflux calculations, and comparisons to techniques, have relied considerably on the methods proposed by the classical Underwood equations for constant relative volatility systems. Numerous methods exist to calculate minimum reflux for non-ideal or azeotropic mixtures; however, these methods require multiple iteration steps. Column Profile Maps (CPM) offer a range of graphical tools to synthesize and analyse the behaviour of simple, as well as complex, distillation configurations. The eigenvectors associated with CPMs have proven to be highly useful for the analysis and, therefore, synthesis of distillation configurations. The eigenvectors for a CPM are a function of thermodynamics only and are thus fixed for a defined system, and do not vary with reflux. Previous research pertaining to the CPM techniques have shown that the eigenvectors of CPM's can be used to non-iteratively determine the minimum reflux requirement for any zeotropic distillation process relating to sloppy split product specifications. Furthermore this method, called the CPM-E technique, can be used to determine the feasible composition region for product specifications given a particular feed. In this work we investigate and extend the use of this the CPM-E technique to more general systems involving non-sharp splits as well as non-ideal thermodynamics. Though the CPM-E technique can calculate the minimum reflux accurately for sloppy split product specifications, without modification to the technique, it becomes a close approximation, not an exact solution, of the minimum reflux for non-sharp split product specification. This is caused by the disappearance of the co-linearity criteria that exists for sloppy splits. This work proposes the use of other geometrical conditions in combination with CPMs to calculate the minimum reflux.

Keywords: Minimum reflux, non-sharp splits, column profile maps, eigenvectors

*Corresponding author contact details:

Email: rabbas@dual-h.com

See more of this Session: Following Dr. Fair's Footsteps: Fractionation and Tray Optimization
See more of this Group/Topical: Topical 8: The Dr. James Fair Heritage Distillation Symposium

On the Selection of Mass-Transfer Internals for CO_2 Capture by Amines

Clint Aichele, Prakash Karpe, and Randy Heald (ConocoPhillips Company), 214187

Aqueous amine absorption is a mature technology for treating natural gas and refinery gas containing CO_2 and H_2S. For that reason, it will likely be the initial technology of choice for capturing CO_2 from flue gas streams for carbon capture and storage. However, the large gas volumes and low pressures associated with flue gas will result in process units much larger and more expensive than conventional acid gas treaters. At the large scale required for CO_2 capture, optimal design of the absorber and regenerator will become important for the minimization of both capital and operating costs. This presentation will focus on the design of the absorber and regenerator for capturing CO_2 from flue gas. Two critical factors for the absorber and regenerator design are the flow parameter and the liquid-film mass-transfer coefficient. The flow parameter is a measure of column liquid loading and pressure for the selected column packing. The flow parameter contributes fundamentally to the selection of column internals and specification of column width. The gas-liquid mass-transfer coefficient is a property of the absorbent that is obtained experimentally. This property contributes to the determination of the absorber height.

(9e) A Novel Approach to Quick-Sizing Tray Towers

Daniel R. Summers, Sulzer Chemtech USA, Inc., Tulsa, OK

There is an interest in the industry to quickly determine tower diameter of conventionally valve trayed towers. The author has modified the heritage "Glitsch Equation 13" to make the determination entirely based on liquid load. The new equation along with adequate downcomer sizing is employed to determine diameter and the number of tray passes for a given tray spacing. A comparison to data is shown and the reasoning behind the recent "Chemical Engineering - Facts at your Finger Tips" article is explained.

See more of this Session: Following Dr. Fair's Footsteps: Fractionation and Tray Optimization
See more of this Group/Topical: Topical 8: The Dr. James Fair Heritage Distillation Symposium

(9f) Optimizing Cross-Flow Tray Performance

Izak Nieuwoudt, Koch-Glitsch, Wichita, KS and Neil A. Sandford, Koch-Glitsch LP, Wichita, KS

Optimizing the performance of a cross-flow tray entails balancing choke, backup and jet flood without losing sight of the mass transfer efficiency. The flood parameters and the efficiency are influenced by the active area and downcomer layout and, very importantly, the bubbling devices. The influence of physical properties and phase ratios are also very important. This multi-dimensional optimization problem requires a large amount of data covering a wide range. Koch-Glitsch has invested in generating a vast body of data that includes data from a large diameter hydrocarbon pilot plant tower. This has allowed the optimization of tray performance. In this paper data will be shown that demonstrates the performance gains that were possible. This will include data from Koch-Glitsch pilot plants and F.R.I. tests. Industrial case studies will also be presented.

See more of this Session: Following Dr. Fair's Footsteps: Fractionation and Tray Optimization
See more of this Group/Topical: Topical 8: The Dr. James Fair Heritage Distillation Symposium

(14a) The Energy Challenge: Providing Sustainable Supplies to Meet Growing Demand

F. Emil Jacobs, ExxonMobil Research and Engineering Company, Annandale, NJ

As economies grow and living standards improve, the world faces a dual energy challenge: meeting the growing demand for energy while mitigating emissions and their impact on the environment. ExxonMobil believes that addressing this dual challenge requires an integrated set of solutions which includes increasing the supply of all commercially viable energy sources, reducing emissions, and improving energy efficiency in all sectors of the economy. Technology will be a key enabler of continued progress in all of these areas.

Many experts agree that the immense scale of global energy demand necessitates a continuing role for oil and gas as a substantial portion of the energy supply mix for at least the next several decades. These experts also forecast that the overall energy mix will shift toward a relatively higher level of sustainability through a combination of increased efficiency across the economy, growth in lower-carbon sources of energy, and development and deployment of technologies that help mitigate emissions.

ExxonMobil has developed a number of technologies that contribute to increased supplies of energy, reduced emissions, and higher efficiency in its operations and in consumer use of its products. The presentation will provide an overview of ongoing research and development activity in a broad range of technology areas including algae-based biofuels. It will also highlight some of company's current technologies including lithium-ion battery separator films, lightweight plastics that improve automotive efficiency, and processes that could advance carbon capture and storage.

See more of this Session: Spring Meeting Keynote Address - Sponsored by the AIChE Fuels & Petrochemicals Division
See more of this Group/Topical: Liaison Functions

(26a) Improving FCC Performance with Low Cost Tower Internals Modifications

Gregory A. Cantley, Technologies, Marathon Petroleum Company LP, Ashland, KY, Austin C. Calico, Robinson Refinery, Marathon Petroleum Company LP, Ashland, KY and Dennis J. Schmude, Koch-Glitsch LP, Houston, TX

A recent revamp of the FCC unit at the Marathon Petroleum Robinson refinery demonstrates the ability to achieve significant improvement in unit performance with relatively low cost tower internals modifications. Proper internals selection and design proved critical to the performance of the main fractionator, primary absorber, and slurry stripper columns.

The major objective of the revamp was to recover lost capacity in the main fractionator due to endpoint changes on product specifications. The undercutting caused a 7.5% decrease in feed rate to the FCC, which was a significant economic impact. The tower was already converted to structured packing in a 1988 revamp, so the recovery of the lost capacity was challenging. Careful analysis of transition zones between beds indicated a problem with an orifice deck distributor. The distributor was replaced along with a change to the packing, which resulted in a 13% increase in feed rate.

During the turnaround, the packing in the primary absorber was replaced with the wrong size due to a documentation error. During troubleshooting of this tower it was discovered that naphtha was being bypassed around the tower due to flooding. The investigation identified a transition zone restriction that when fixed added an additional 7% of propane recovery and 11% of propylene recovery.

A new slurry stripper column was added to increase the flash of the slurry going to storage. Special fouling resistant trays were installed due to the concern for fouling. The tower operation has been trouble-free and has been successful in meeting the desired flash point specifications.

See more of this Session: Improved Design and Operation in Refinery Distillation
See more of this Group/Topical: Topical 8: The Dr. James Fair Heritage Distillation Symposium

(26b) Hydrocarbon Strippers - the Importance of Water Solubility

Karl-Martin Hansson, Operations, Preem AB, Lysekil, Sweden

Many control room operators and design engineers assume that stripping steam added to a distillation column, will infallibly end up in the column overhead. This is normally not the case, especially not for columns operating at high pressures. Since the water flow from overhead reflux drums is rarely measured, there is the chance that a column can run for an extended period of time with no steam reaching the column top, without anyone noticing.

There are two situations when no stripping steam goes through the column and hence, the added stripping steam does not do any stripping, but does only wet the bottom product. The first situation is when the amount of stripping steam used is less than the amount of steam that can be dissolved in the hydrocarbons in the bottom of the column. This situation might seem a bit unrealistic, but for all three high pressure strippers (5,8 barg or higher) in one refinery, 50% or more of the stripping steam used, was dissolved in the hydrocarbons in the bottom of the columns. There have even been incidents in the past where all or nearly all of the stripping steam ended up in the bottom of the column due to this fact. For one of the columns, the amount of stripping steam that dissolved in the bottom hydrocarbons, was more than 75% of the design flow rate of stripping steam.

The second situation that can render stripping steam useless, is when the inlet temperature to the column is lower than the condensation temperature for steam at the column bottom pressure. This situation might seem even more unrealistic. Even so, a column in a HDS unit running in block mode, switching back and forth between processing kerosene and light gasoil every week, had complete condensation of all added stripping steam every time the column was running in kerosene mode, for at least 10 years. This led to very poor stripping of light gases in the column, which led to excessive flaring from a downstream fractionator every time the unit was running in kerosene mode. Since excessive flaring is not allowed, this limited the unit throughput. Increasing the column inlet temperature by about 15 °C was enough to avoid that the stripping steam condensed and the flaring problem disappeared.

Many simulation tools using equation of state models to estimate vapour-liquid equilibrium, have problems estimating three phase equilibrium when water is present. A simple equation for maximum water solubility in hydrocarbons has been suggested in literature. The equation states that the water solubility does only depend on temperature, pressure and average molecular weight for the hydrocarbon. That equation was compared with the observed solubility in three columns with oil fractions with very different molecular weights. The estimated solubilities are in good agreement for all fractions, and can be used to check the simulation results from more rigorous models in the design of new units.

See more of this Session: Improved Design and Operation in Refinery Distillation

(26c) Innovative Stripper Revamp Improves Hydrocracker Operation

Daryl Hanson, Valero Energy Corporation, San Antonio, TX and Patrick Williams, Valero Port Arthur Refinery, Port Arthur, TX

Valero Port Arthur refinery operates a 40 MBD Hydrocracker that has a front-end stripper. After the passing through the conversion reactors, the Stripper removes the light hydrocarbons (butane and lighter, including Hydrogen Sulfide) from the bottoms product. The bottoms product is further processed into finished blendstocks in the downstream Product Fractionator.

The original Stripper design included packed sections above and below the feed zone. The Stripper never operated well and prevented the Fractionator Light Naphtha from meeting specifications. Feed zone distributor and packing fouling and subsequent flooding was identified as problem areas. This limited Hydrocracker rate and conversion and frequently resulted in refinery slowdowns due to Stripper liquid carryover.

During a 2009 turnaround, the bottom of the Stripper was revamped from two beds of packing to trays. The revamp utilized a novel support system that stacked approximately 10-12 trays on a single support ring, and did not require welding to the tower shell. The two-pass trays were supported from surrogate rings and support pedestals which were sandwiched together. Packing removal and tray installation was accomplished in 8.5 days, which was only slightly longer than the previous packing replacement turnaround duration.

After almost a year of operation, post-revamp data indicates that the trays are performing much better than the packing. There is no evidence of fouling degradation, the Fractionator Light Naphtha sulfur content has decreased from 90-100 ppm to 5-10 ppm, and Stripper flooding has been eliminated. The authors will present pre-revamp data, describe the revamp strategy, and present post-revamp data that will show that

trays were the best choice for this high liquid rate/low vapor rate service.

See more of this Session: Improved Design and Operation in Refinery Distillation
See more of this Group/Topical: Topical 8: The Dr. James Fair Heritage Distillation Symposium

SULZER
Sulzer Chemtech

The new Schoepentoeter* Plus
A step ahead in the bulk separation of gas-liquid mixtures

By

Giuseppe Mosca, Pierre Schaeffer, Bart Griepsma, Sulzer Chemtech

Presented at the

AIChE Spring National Meeting
14-17 March 2011, Chicago, IL

(*) Schoepentoeter is a trademark owned by Shell

ABSTRACT

In the hydrocarbon industry—i.e., oil refineries, petrochemical plants, gas plants, and exploration and production facilities—it is often necessary to separate gas from liquid at a certain stage of an operation or process. The operating conditions of the mixed phases and the requirements for separation efficiency may vary widely. Therefore, special care should be taken in selecting the most appropriate device to match the specific duty.

For a bulk gas-liquid separation, where generally not more than 95% of the liquid must be removed from the gas stream, the Schoepentoeter has proven to be one of the most effective devices available on the market. The Schoepentoeter is a Shell proprietary feed inlet vane device commonly used for introducing gas-liquid mixtures into distillation columns or gas-liquid separators.

In most of the cases, the conventional Schoepentoeter has been proven to provide very good performance, and even to exceed expectations.
There are indeed only a few applications, e.g., refinery vacuum towers, where the separation efficiency was measured to be lower than expected.

To comply also with the most severe services, Sulzer Chemtech and Shell Global Solutions have jointly developed the new Schoepentoeter Plus which substantially improves the performances of the conventional one.

(26e) Experiences with the Use of Gamma Scanning

German A. Luna Sr., HSE, ACM FACILITY SAFETY, Calgary, AB, Canada

The Gamma Scanning technique was developed many years ago; and as in many other inventions or discoveries, its misuse is really a problem and can have a strong impact in money and time. Doing a proper gamma scanning requires a detailed plan and strategy, which will be discussed in this paper. Bad results are normally related with the misuse or misinterpretation of the technique. The intent of this paper is to present our experiences and strategies to perform a good gamma scanning. To reach that point we also will present our experience with bad and good results, including specific examples in amine towers, crude towers and vacuum towers.

See more of this Session: Improved Design and Operation in Refinery Distillation
See more of this Group/Topical: Topical 8: The Dr. James Fair Heritage Distillation Symposium

JACOBS Consultancy

BTX Distillation: Conventional, Pressure Cascade, or Dividing Wall?

Dennis E. O'Brien
Laura E. Weaver

Jacobs Consultancy
Contact: laura.e.weaver@jacobs.com

Presented at the Dr. James Fair Heritage Distillation Symposium

AIChE Spring Meeting

Chicago, IL

March 13-17, 2011

BTX Distillation Options

- Conventional two-column distillation
- Pressure cascade two-column distillation
- Dividing wall column distillation

→Dividing wall columns save money on capital costs vs. both alternatives, and often reduce operating costs and emissions as well

JACOBS Consultancy

The Need for Energy Efficiency

- Increase in utility costs
  - Units currently operating were designed for lower energy costs; suboptimal energy usage
- Greenhouse gas regulations
  - Mandatory emissions reduction
  - Even higher effective fuel cost

Conventional Two-Stage Distillation

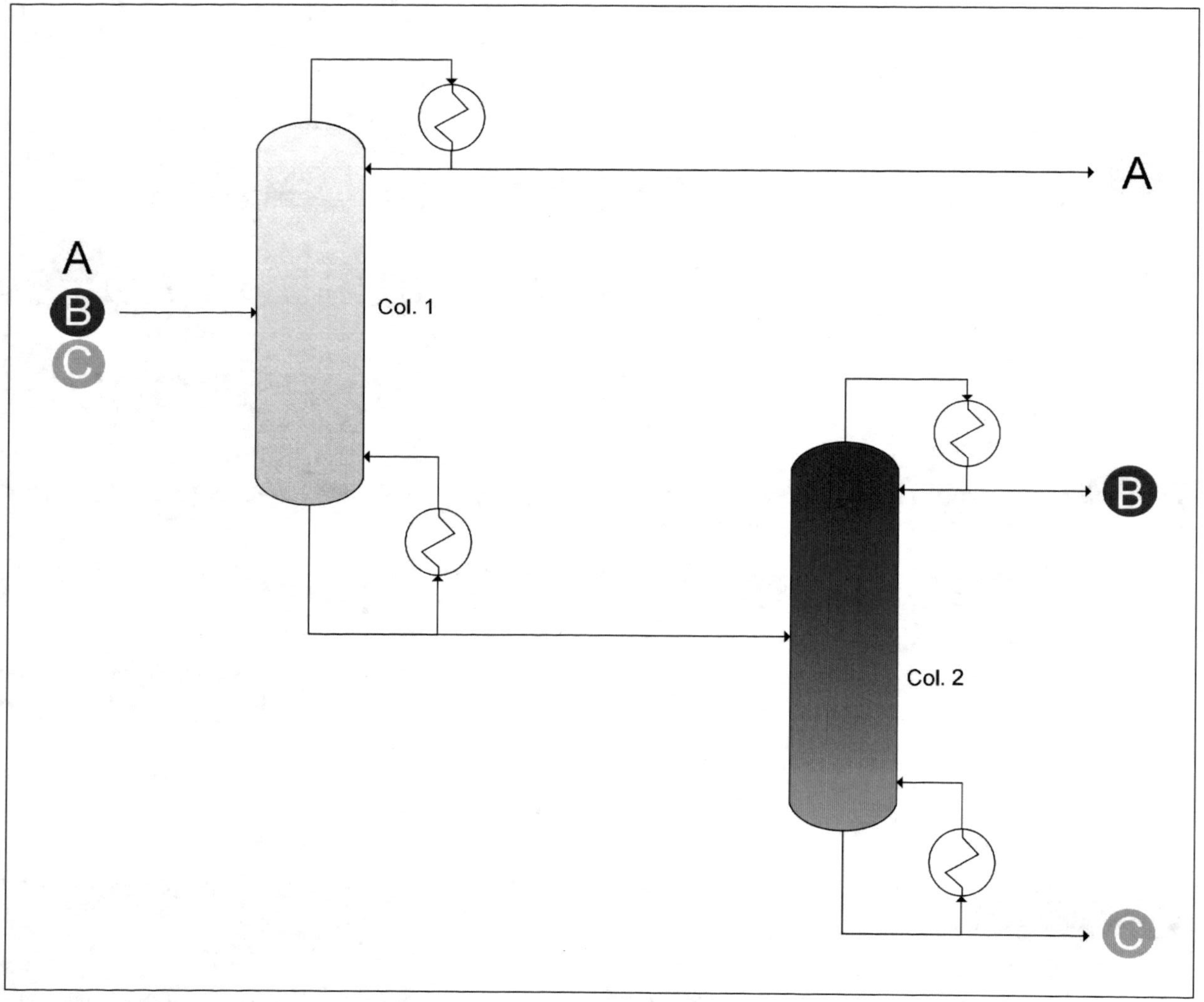

JACOBS Consultancy

Conventional Distillation

- Normal sequence for multicomponent distillation
  - Lift lightest product first
  - Continue to the next lightest product
- This minimizes the number of vaporization and condensation steps

JACOBS Consultancy

Pressure Cascade Distillation

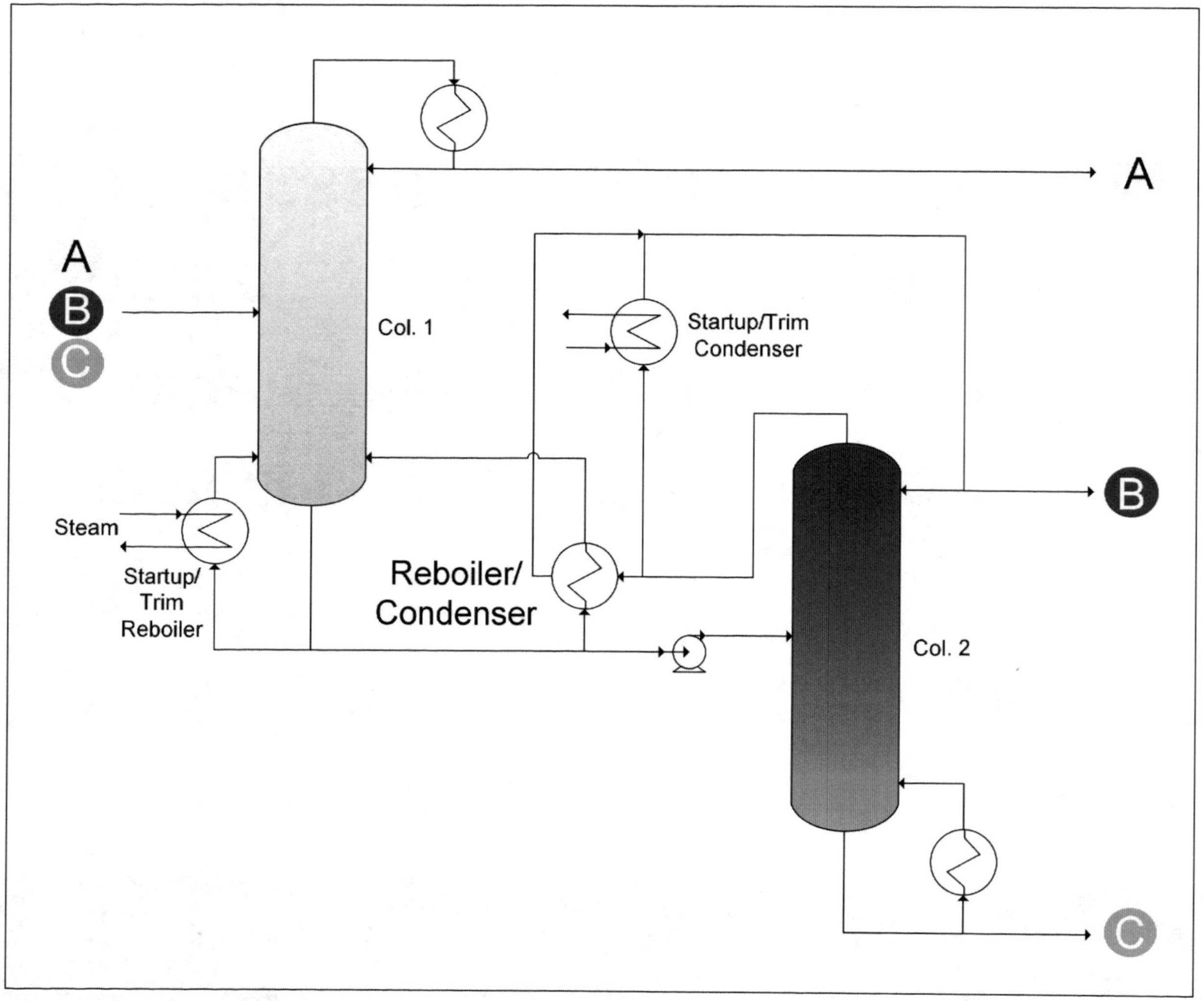

JACOBS Consultancy

Pressure Cascade Distillation

- Second column is run at slightly increased pressure
- First column is reboiled with the overhead of the second
- *This reuses the energy to separate B from C to separate A from B+C*
- Various configurations of pressure cascade exist

JACOBS Consultancy

Dividing Wall Distillation

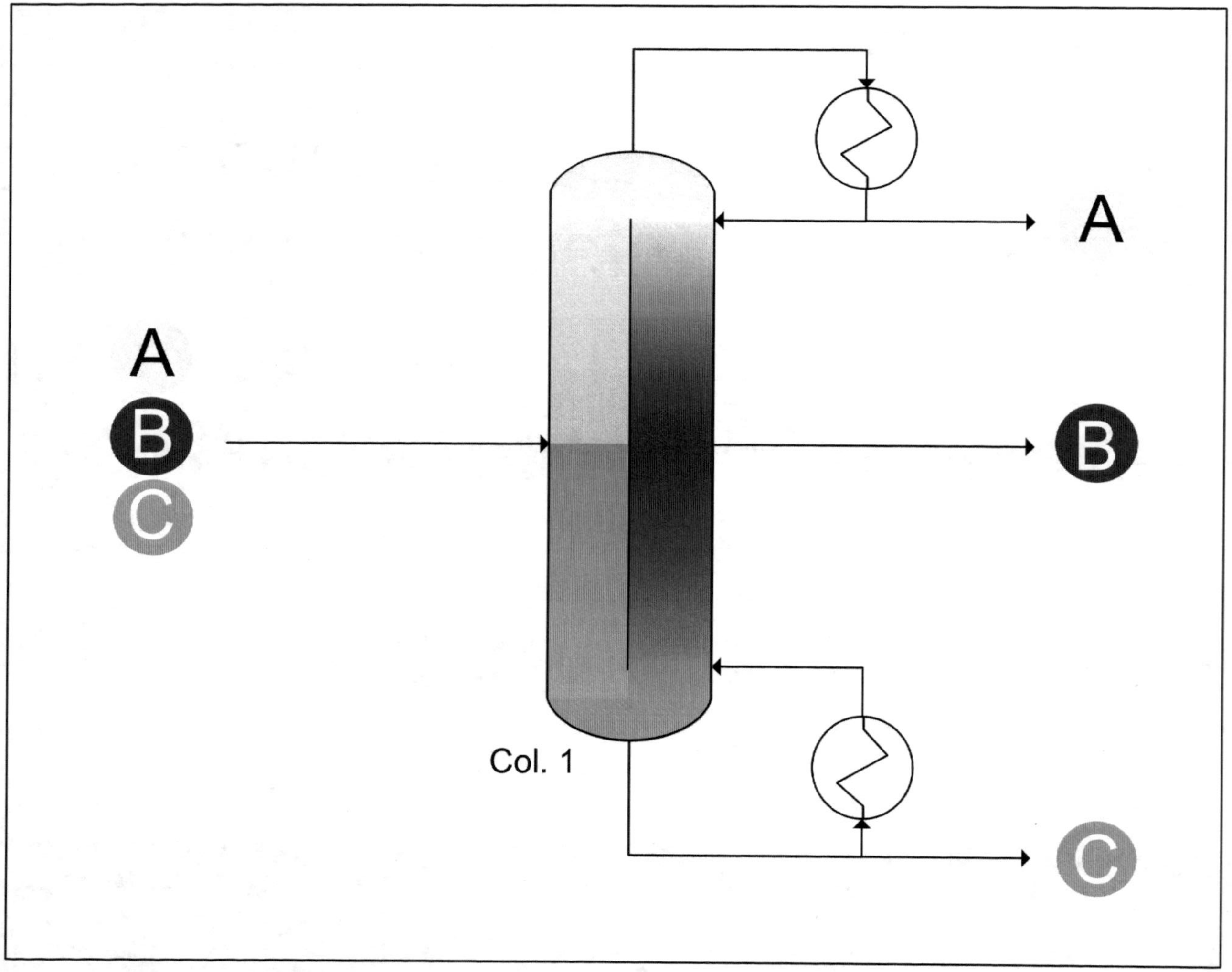

JACOBS Consultancy

Dividing Wall Distillation

- Combine both columns into a single column with a dividing wall
- *The energy used to separate B from C is reused to separate A from B in the same shell*

Case 1: BTX Distillation, 2010 Work

- Jacobs simulated a typical benzene-toluene-xylene fractionation
  - Conventional two-column
  - Pressure cascade
  - Dividing wall column
- Jacobs developed a method for modeling the dividing wall column with the others
  - Aspen HYSYS was used for process simulation, with verification done in Aspen Plus
  - Peng-Robinson thermo package was used

BTX Utility Requirements: 2010 Work

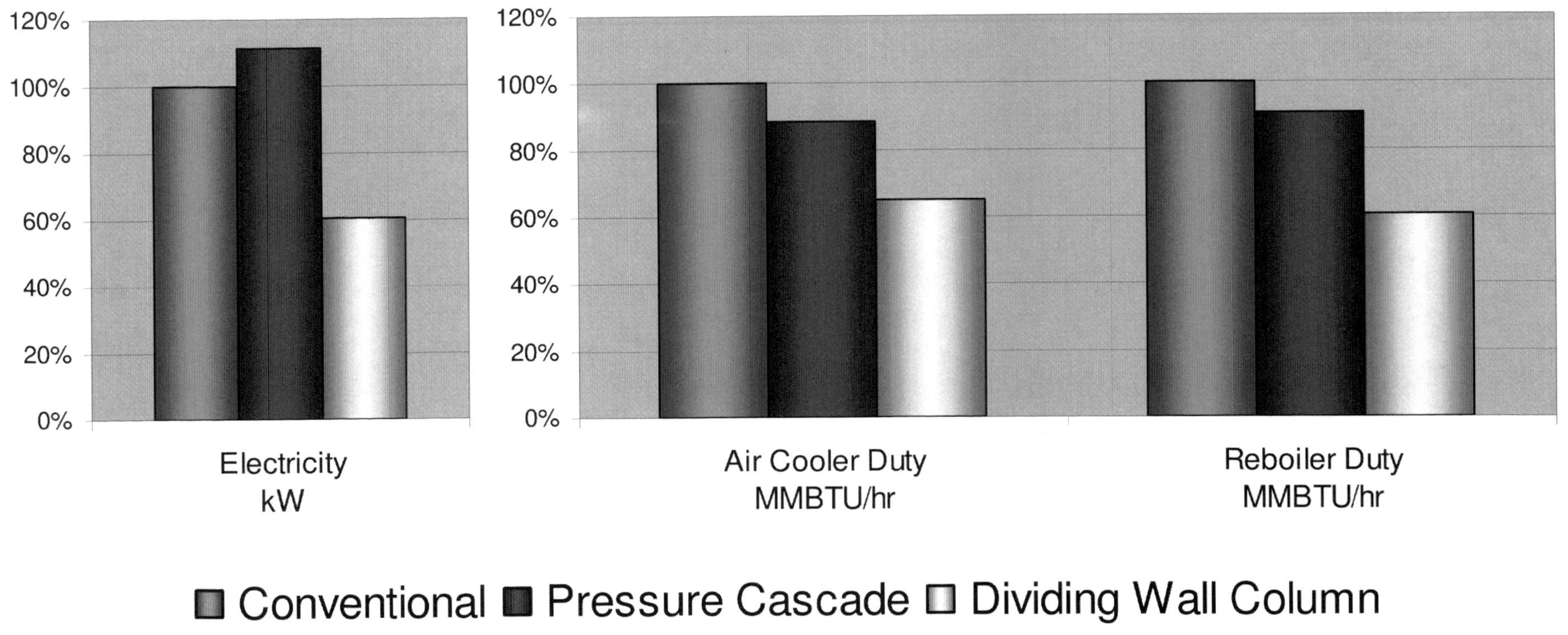

JACOBS Consultancy

BTX Equipment Count: 2010 Work

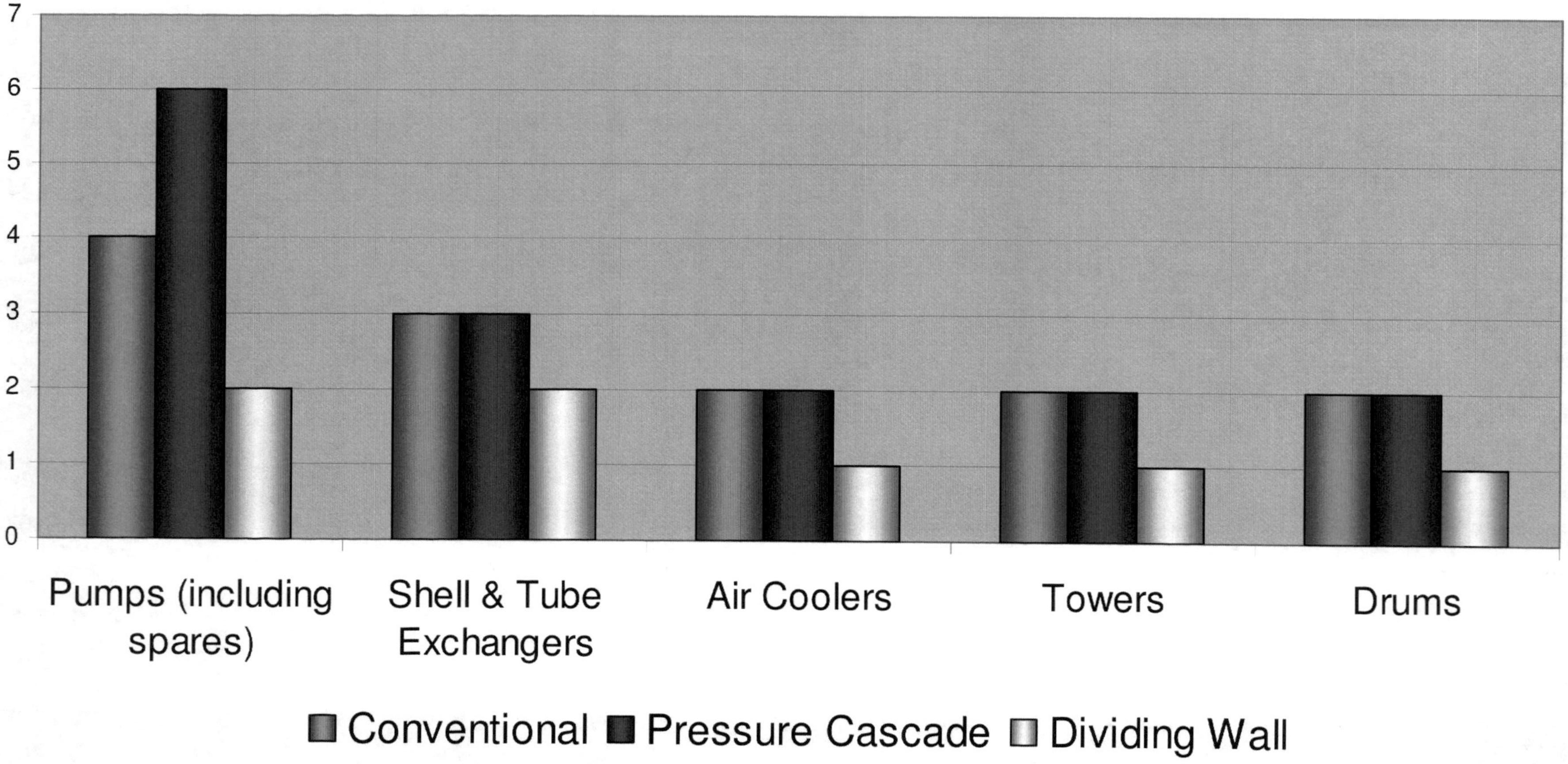

JACOBS Consultancy

BTX Capex/Opex: 2010 Work

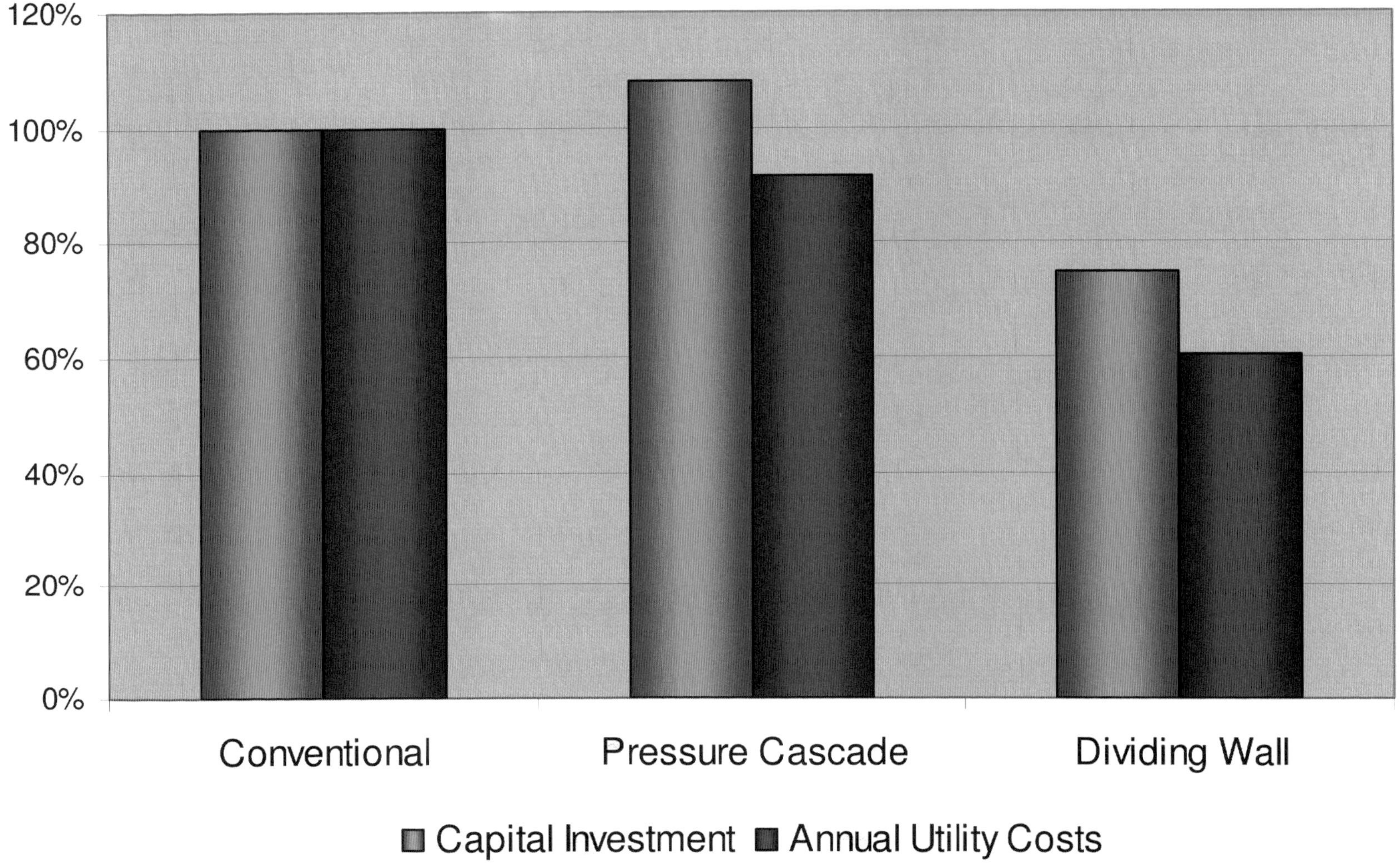

JACOBS Consultancy

Case 2: BTX Distillation, 2011 Work

- As before, Jacobs simulated a typical benzene-toluene-xylene fractionation
  - Conventional two-column
  - Pressure cascade
  - Dividing wall column
- Fractionation updated to meet 545 Benzene and Nitration-Grade Toluene ASTM specifications
- Column pressure increased to 10 psig overhead to avoid contamination from flare header backflow, per recent design methods
- Tray count increased to accommodate more stringent specifications

JACOBS Consultancy

BTX Utility Requirements: 2011 Work

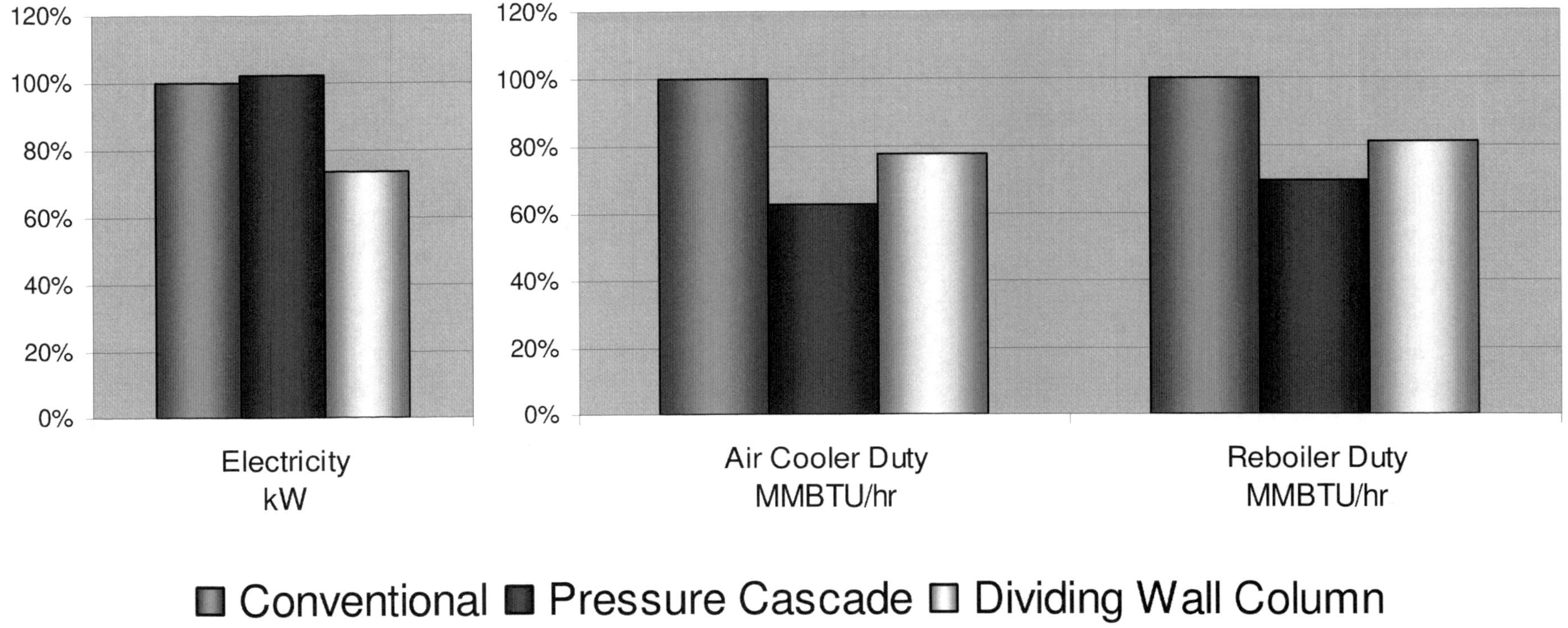

JACOBS Consultancy

BTX Steam Requirements: 2011 Work

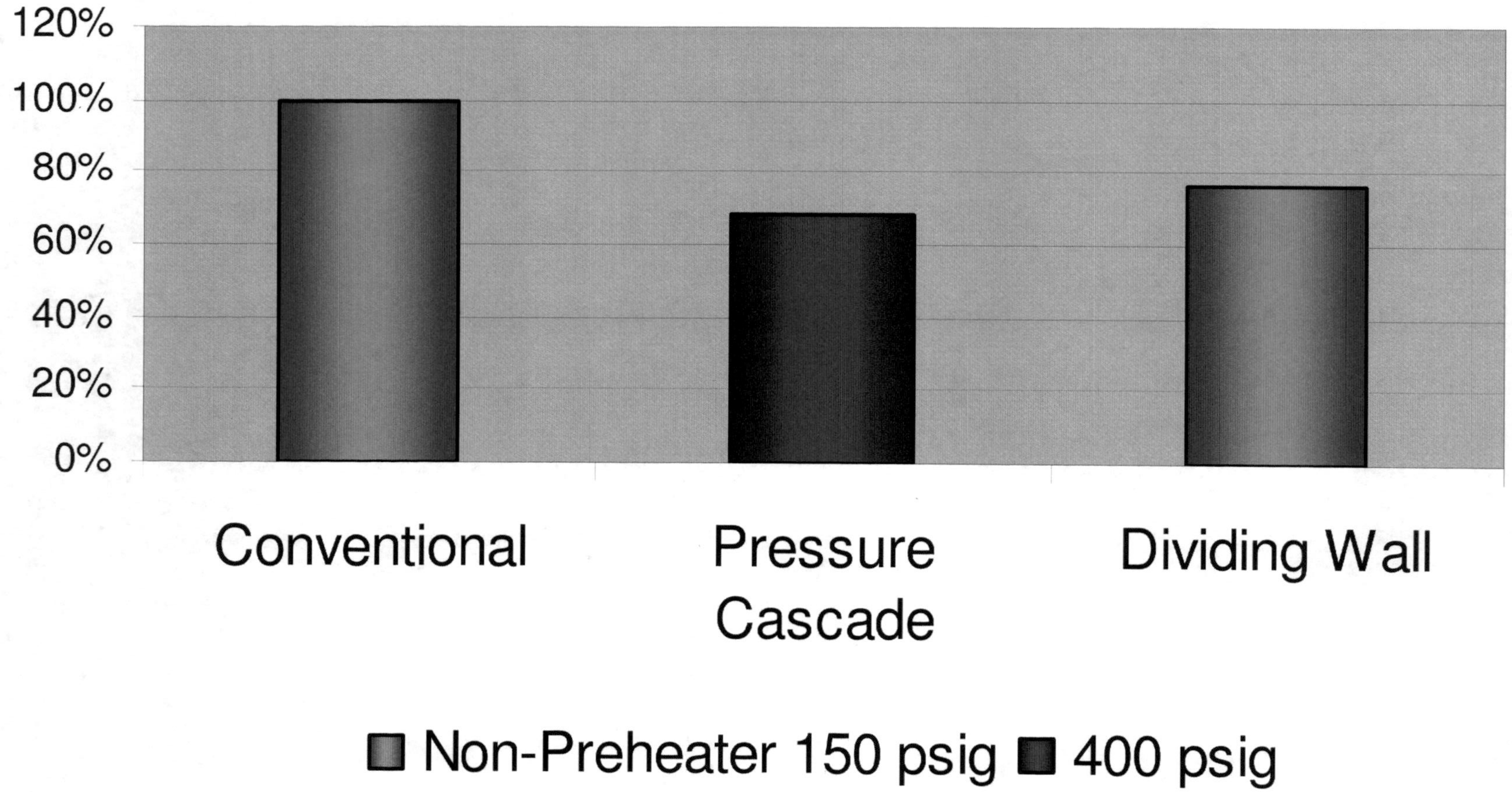

Steam requirements are displayed in mass units, not energy units.

Note: The preheater duty, with 150 psig steam, is equal for each case, and would add another 25% to the conventional case.

JACOBS Consultancy

BTX Equipment Count: 2011 Work

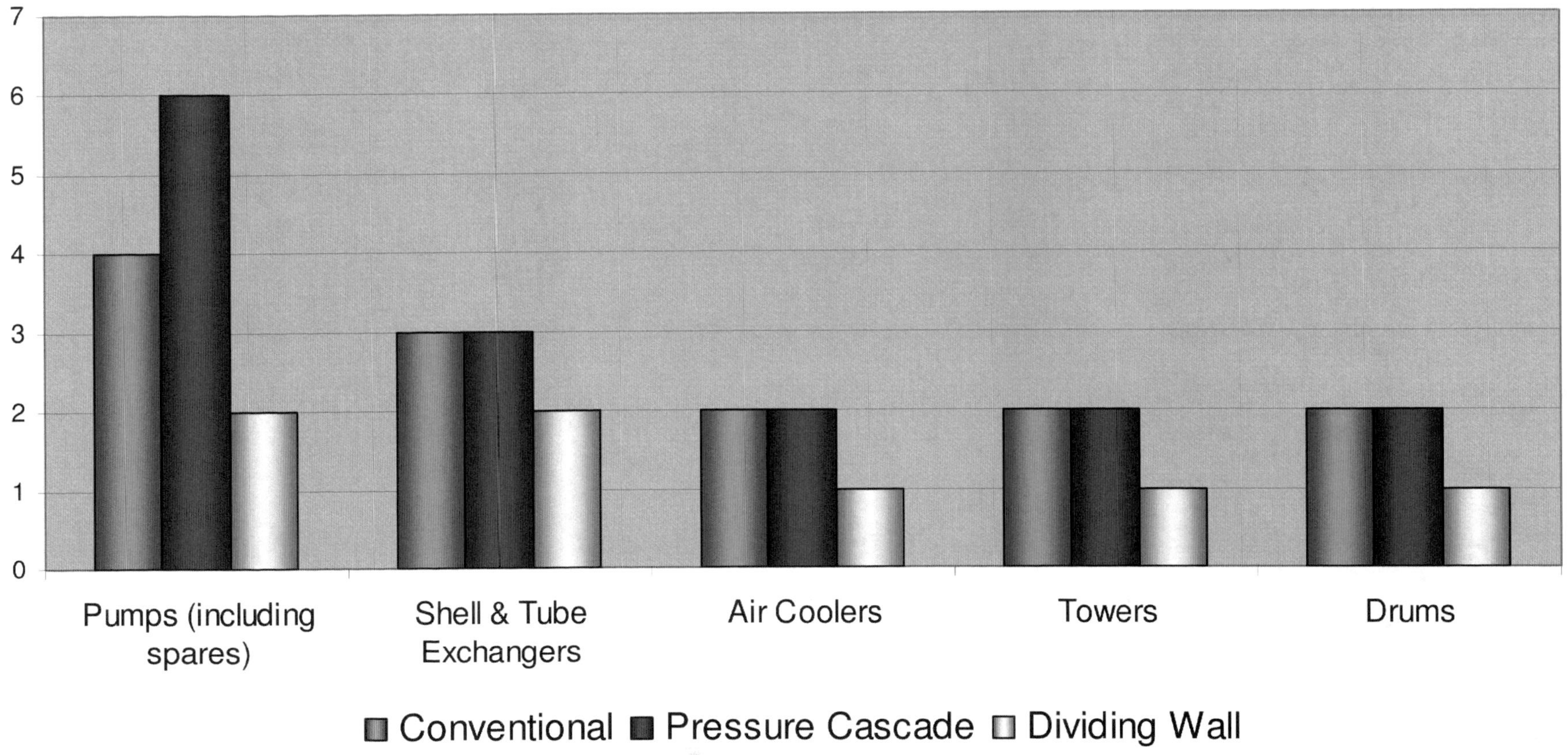

JACOBS Consultancy

BTX Capex/Opex: 2011 Work

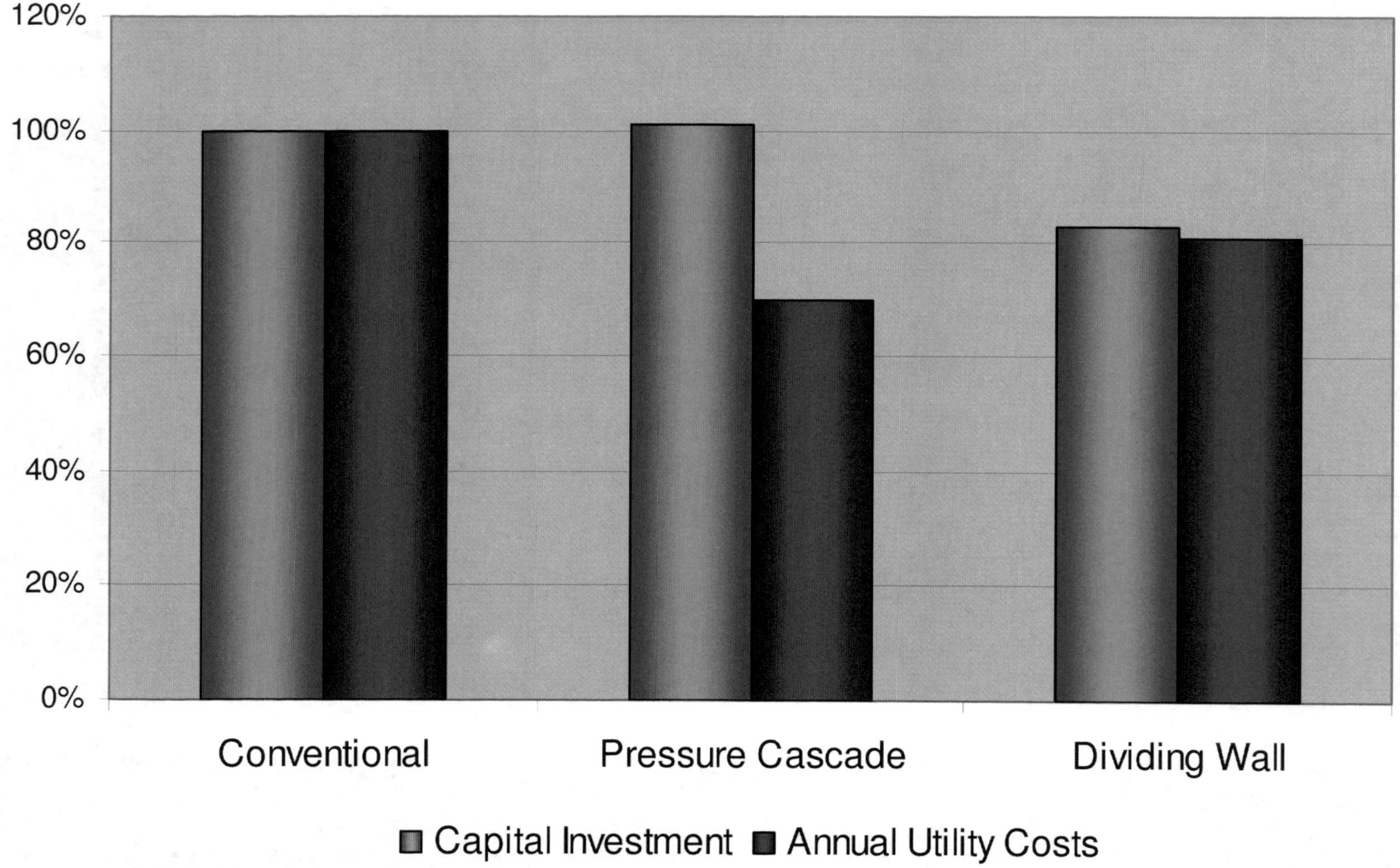

JACOBS Consultancy

BTX DWC Distillation

- More severe benzene-toluene split required by the more rigorous specifications
- Required air cooler and reboiler duties for the DWC increased with the higher specifications

- However, DWC system still shows clear savings in capital costs over both alternatives, and in operating costs over the conventional system
- DWC may also require lower-quality heat than pressure cascade

BTX Distillation Results

- Pressure Cascade configuration:
  - Similar capital investment requirement to conventional two-column distillation
  - Modest-to-significant improvement in utility costs
  - Existing conventional systems may be revamped to pressure cascade
- Dividing Wall Column configuration:
  - Provides significant savings for both capital and operating costs
  - Requires smaller plot space
  - Design of control scheme is critical

JACOBS Consultancy

Conclusions

- Conventional scheme is more costly and less efficient
- Pressure cascade can be more energy efficient
- Dividing wall is more capital efficient and energy efficient

JACOBS Consultancy

Future Perspectives of Using Hollow Fibers as Structured Packings in Light Hydrocarbon Distillation

Dali Yang (dyang@lanl.gov, 505-665-4054), Loan Le, Ronald Martinez, Bruce Orler, Cindy Welch, and Stephanie Tornga
Los Alamos National Laboratory, Los Alamos, NM 87545, U.S.A

Introduction

Olefins and paraffins are the largest volume chemical commodities. Furthermore, they are major building blocks for the petrochemical industry. Each year, petroleum refining, consumes 4,500 TBtu/yr in separation energy, making it one of the most energy-intensive industries in the United States[1]. Just considering liquefied petroleum gas (ethane/propane/butane) and olefins (ethylene and propylene) alone, the distillation energy consumption is about 400 TBtu/yr in the US. Since petroleum distillation is a mature technology, incremental improvements in column/tray design will only provide a few percent improvements in the performance. However, each percent saving in net energy use amounts to savings of 10 TBtu/yr and reduces CO_2 emissions by 0.2 MTon/yr[2,3]. In practice, distillation columns require 100 to 200 trays to achieve the desired separation. The height of a transfer unit (HTU) of conventional packings is typical in the range of 36 – 60 inch[4]. Since 2006, we have explored using several non-selective membranes as the structured packings to replace the conventional packing materials used in propane and propylene distillation. We obtained the lowest HTU of < 8 inch for the hollow fiber column[5-7], which was >5 times shorter than that of the conventional packing materials. In 2008, we also investigated this type of packing materials in iso-/n-butane distillation[8]. Because of a slightly larger relative volatility of iso-/n-butane than that of propane/propylene, a wider and a more stable operational range was obtained for the iso-/n-butane pair. However, all of the experiments were conducted on a small scale with flowrate of <25 gram/min. Recently, we demonstrated this technology on a larger scale (<250 gram/min). The high separation efficiency is obtained for the n-/iso-butane distillation under either total and changeable reflux conditions. Up to 35% enrichment of iso-butane is achieved in a hollow fiber column with a < 37 inch length. Within the loading range of F-factor < 2.2 $Pa^{0.5}$, a pressure drop on the vapor side is below 20 mbar/m. These results suggest that the pressure drop of hollow fiber packings is not an engineering barrier for applications in distillation. The thermal stability study suggests that polypropylene hollow fibers are stable after a long time exposure to $C_2 - C_4$ mixtures. The effects of packing density on the separation efficiency will be discussed.

Experimental

In this work, we have constructed a new distillation apparatus, which can be operated at the flowrate up to 250 gram/min. The new apparatus has 10 times the capacity of the previous system[5, 8]. The apparatus was comprised of reboiler, hollow fiber column, condenser, heat exchanger, circulation pump, auto-sampling system, catch tank, four mass flowmeters, two pressure differential pressure gauges, five pressure gauges, and many thermal couples. The operational parameters were recorded by Labview program in real time. The auto-sampling system was controlled by Labview program as well. A certain amount of iso-/n-butane mixture was fed into the reboiler. By controlling the temperature setting at the reboiler or/and the condenser, we could to control the process flowrate and operational temperature in the experiments. When we divided some portion of liquid into the catch tank, we could mimic the operations under the changeable reflux condition in addition to the total reflux condition. A circulation pump was added between the condenser and the membrane column, which ensured the liquid flow direction. The mixture of iso-/n-butane (50%/50%) was purchased from Trigas (Irving, TX). Ultra-pure helium and argon (>99.99%) were used as Micro-GC mobile phases. Two standards (propylene/propane and iso-/n-butane mixtures) were used to calibrate the micro-GC (Agilent).

Hollow modules were made of Celgard® polypropylene (PP) hollow fiber with 240 μm ID and 30 μm wall. The fiber porosity was ~ 40%. The dimensions of module (length x ID) was ~ 90 cm x 2.54 cm. To investigate the effect of the packing density on the module performance, the packing density varied from 5.5% to 18.5%.

To evaluate the thermal stability of the PP hollow fibers, the fibers were immersed in iso-/n-butane mixture, benzene, and cyclohexane at 50 °C for different times. The thermal gravimetric analysis (TGA), differential scanning calorimetry (DSC), dynamic mechanical analysis (DMA), scanning electron microscopy (SEM) and Brunauer, Emmett, and Teller (BET) were performed on the aged samples to determine the effect of solvent exposure on the thermal, mechanical and morphological properties of the hollow fibers.

Results and Discussion

1. Capacity of hollow fiber modules

In the olefin/paraffin systems, (e.g. ethane/ethylene, propane/propylene, and n-butane/iso-butane), flooding is commonly encountered when the column is operated at a high capacity. This is because the columns are operated under conditions where the gas and liquid densities are similar. Therefore, the flooding is a more severe restriction. Therefore, lack of flooding is especially important for the new type of structured packing materials when they can be used at the high capacity condition. Generally, the capacity of the packing materials can be determined by the correlation between flow parameter and capacity factor. The flow parameter can be calculated using:

$$\text{Flow parameter} = \left(\frac{L'}{G'}\right)\left(\frac{\rho_G'}{\rho_L'}\right)^{0.5}, \quad (1)$$

where L' and G' are the liquid and gas mass fluxes (lb/s ft^2), respectively, and ρ_L' and ρ_G' are the liquid and vapor density (lb/ft^3), respectively. The capacity factor depends on the ratio of the kinetic energy in the vapor to the potential energy in the liquid, which is defined as:

$$\text{Capacity factor} = \frac{G'^2 F \psi \eta^{0.2}}{\rho_G' \rho_L' g_C} \quad . \quad (2)$$

where F is the packing factor (ft^{-1}), ψ is the density ratio between water and process liquid, η is the viscosity (cP) of process liquid, g_C is the gravitational constant (32.2 lb-ft/lb$_f$-sec^2).

Capacity factors and flow parameters for the five hollow fiber modules are plotted in Figure 1, where the solid line gives the upper-limit capacity of conventional packing materials. However, in the hollow fiber packing both liquid and vapor have their own channels in which to flow, so the flooding encountered in a conventional packed tower is circumvented. Thus most of our experimental points lie above this solid flooding line. This means that hollow fiber modules can be over at least 10 times more productive per unit volume than conventional packing.

In general, the flow parameter is related to the cross-sectional area ratio of vapor to liquid phase. Lower packing density results in large A_G/A_L ratio and hence large L/G. Among the modules tested, LANL 17 has the lowest packing density, and thus has the highest flow parameter.

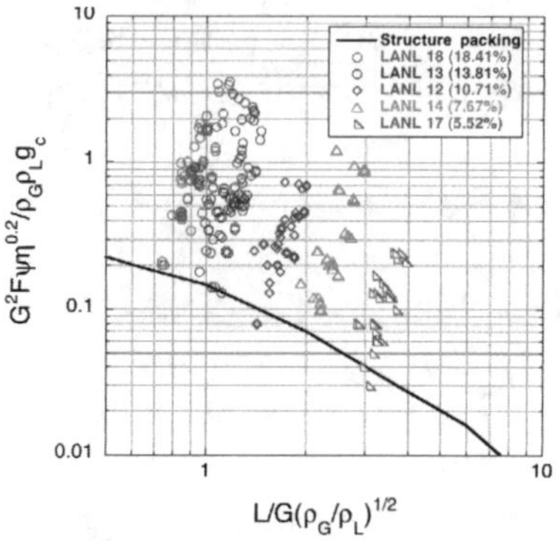

Figure 1. The correlation of flow parameter *vs.* capacity factor for five hollow fiber modules (the operation conditions range from 15 – 40 °C, and 30 – 80 psig for n-/iso-butane system).

2. Effect of packing density on separation efficiency and operational stability

From our study, we have found that the module design significantly impacts the separation efficiency and operational stability of the hollow fiber modules. To measure the separation efficiency, the height of transfer unit (HTU) is commonly used for the distillation columns. The smaller the HTU is the better the separation is. Due to the severe flooding problems, structured packings have not been used in the n-/iso-butane distillations. Instead staged plates are commonly used. The design HTU in those columns is equivalent to distance between two plates, which ranges from 36–60 inch (>90 cm). For the HTU of the packing columns, it can be calculated using:

$$HTU = \frac{V_G}{K * a}, \qquad (3)$$

where V_G is superficial vapor velocity (cm/sec), K is overall mass transfer coefficient (cm/sec), and a is specific area of the column (cm^2/cm^3). For the flow characterization, a commonly used term is the F-factor (F_G) ($(m/s(kg/m^3))^{0.5}/sec$), which can calculated using the superficial vapor velocity at the shell side of the column, and is related to the C-factor (C_G) (cm/sec) through the square root of the difference of liquid and vapor densities:

$$F_G = V_G\sqrt{\rho_G} \quad \text{and} \quad C_G = \frac{F_G}{\sqrt{\rho_L - \rho_G}}. \qquad (4)$$

The effect of packing density on the separation efficiency (HTU) of five LANL modules is shown in Figure 2(a). Among these five modules, LANL 17 has the lowest packing density (5.52%), and gives the worst performance. Due to the smallest cross-sectional area for the liquid phase in LANL 17, the liquid velocity is the highest when the gas velocity is the same among five modules. In the overall mass transfer process, the major resistance comes from the liquid side. The retention time of liquid phase in the module is more critical than that for the vapor phase. For the low packing density, the high liquid velocity shortens the retention time, which reduces the contact time with the vapor phase and lowers the separation efficiency. Furthermore, high liquid velocities result in a large pressure drop in the liquid side for the same F-factor, as shown in Figure 2(b). In a stable operation, the pressure distribution along the hollow fiber wall between two phases is well controlled. The micro-porous wall of the hollow fibers is wetted by the process fluid and holds this thin layer of liquid inside wall of the hollow fibers. Therefore, both liquid and vapor liquid phases have their own flow channels. Although the liquid pressure may be slightly higher than the vapor pressure at the top portion of the module where the liquid is fed into the module after the condenser, the vapor pressure is generally greater than the liquid pressure for the majority part of module. The problem of liquid leaking from the lumen side to the shell side of the module is insignificant. However, when the liquid pressure drop increases appreciably, the portion of the module with the liquid pressure exceeding the vapor pressure will increase. As an appreciable amount of liquid is pushed into the vapor side, it will interrupt the vapor flow and decrease stability of the operation. This reason may explain that why LANL 17 has the least stable zone for the effective separation.

As the packing density increases, the HTU decreases giving increased separation efficiency and a larger stable operating window. An explanation for this observation is the following: as the packing density increases, the cross-sectional area of the liquid phase increases. Therefore, the liquid velocity decreases, which gives a longer retention time for liquid to contact the vapor phase and hence increases the mass transfer between two phases. Simultaneously, the low liquid velocity also results in a lower pressure drop, as shown in Figure 2(b), and decreasing the likelihood of liquid being pushed into the vapor phase. All of these changes allows for better pressure balance along the module and increases operational stability. Therefore, the high separation efficiency is obtained. For both LANL 18 and LANL 13, the HTUs of less than 30 cm are obtained when the capacity is increased to 1.0 $Pa^{0.5}$. The high separation efficiency is obtained for the n-/iso-butane distillation under both total and changeable reflux conditions. Up to 35% enrichment of iso-butane is achieved in LANL 18 module. One may expect that when the packing density increases further, higher separations (HTU < 20 cm) and wider stable zone can be achieved. However, there will be an upper-limit of the packing density. Although high packing density (up to 30%) of hollow fiber modules is commonly used in gas separation and water purification, it is not clear what the practical upper-limit for the hollow fiber modules will be for this application. Over the test range considered here (packing density up to 18.4%), appreciable increases in the vapor pressure drop were not seen. When the higher packing densities begin to appreciably increase the pressure drop on the vapor side, the upper-limit of the packing density may have been approached.

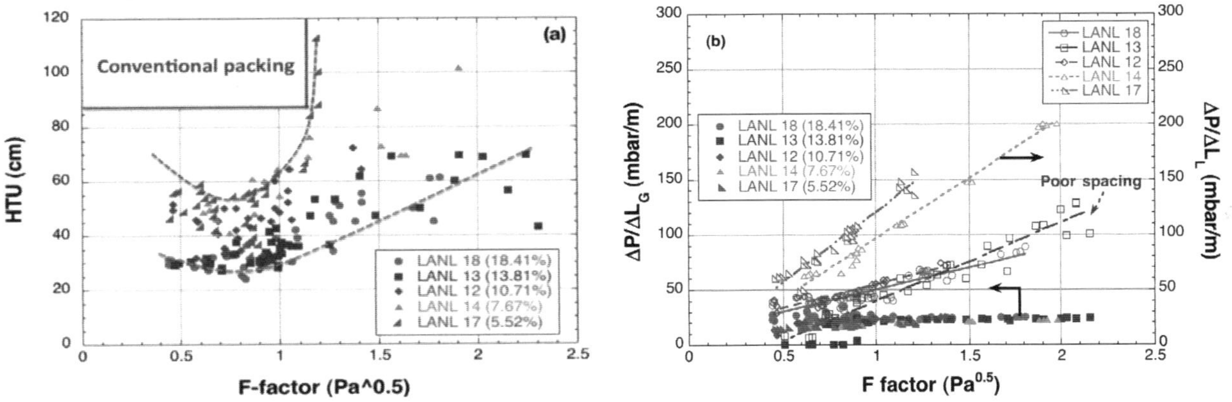

Figure 2. (a) Comparison of separation efficiency among five LANL modules and (b) The correlation between F-factor *vs.* pressure drop in both phases (left y-axis is the pressure drop in vapor side while right y-axis is the pressure drop in liquid side) (the operation conditions range from 15 – 40 °C, and 30 – 80 psig for n-/iso-butane system).

In Figure 2(b), we correlate the pressure drop in both phases *vs.* the F-factor for these five modules. As the packing density increases, the pressure drop in the liquid phase ($\Delta P/\Delta L_L$) largely decreases at the same flow capacity, and also becomes less sensitive to the increased F-factor, which will increase the operational stability. We had expected that the pressure drop in the vapor phase ($\Delta P/\Delta L_G$) would increase as the packing density increased. The experimental results suggest that the $\Delta P/\Delta L_G$ does not notably increase as the packing density increases up to 18.4%. Furthermore, the values of $\Delta P/\Delta L_G$ are below 20 mbar/m, which are very important results. It addresses one of several engineering questions about this type of packing used in distillation applications. The pressure drop is not an engineering barrier for this type of packing materials used in olefin/paraffin distillation[9]. Interestingly, under changeable reflux conditions, the distillation experiments are not only more stable, but also give better separation compared to those obtained under the total reflux condition. An explanation to this may be the less amount of liquid traveling through the lumen side of the module reduces the likelihood of liquid leakage into the shell side, which increases the operational stability, and hence increases the separation efficiency.

3. Thermal stability study

One of the concerns about using polymeric materials as the structured packing is their durability when they are continually exposed to organic solvents, especially at elevated temperatures. The hollow fibers must maintain their mechanical integrity and morphology for the lifetime of the hollow fiber modules. Therefore, we have conducted thermal stability study of the polypropylene (PP) hollow fibers. We have aged the fibers in several organic solvents at elevated temperatures for up to 24 months. In Figure 3, we summarize the TGA results for the PP soaked in benzene and cyclohexane at 50 °C up to 24 months (a) and in iso-/n-butane under distillation conditions (b). The TGA results suggest that PP thermally decompose above 350 °C. Although the organic solvent exposure does not to change their thermal stability appreciably, PP fibers are not recommended to use at temperature above 100 °C.

The TGA results show a slight weight gain before the PP decomposition instead of a weight loss, which is not commonly observed. This weight gain may be due to experimental error. The fine PP fibers experience static repulsion, which makes loading sufficient quantities of sample (>1mg) into a TGA pan difficult. The slight weight gains during heating are most likely due to convection within the TGA furnace. When small samples are used, the convection current from heating gives the false impression of weight gain. Nevertheless, the TGA behavior of PP fibers does not significantly change upon the solvent exposure process.

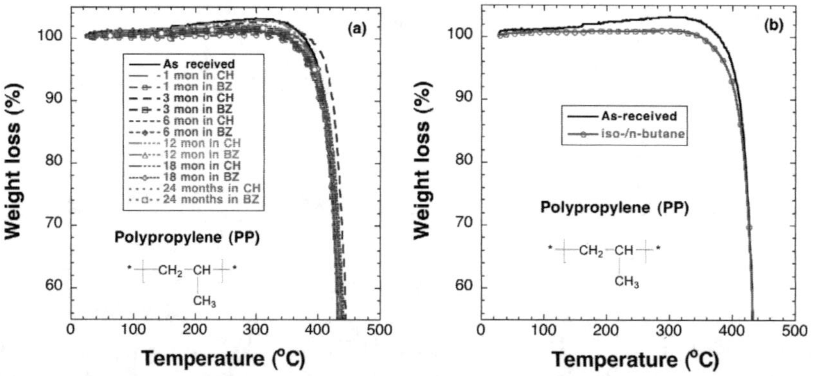

Figure 3. The TGA results for PP fibers immersed in benzene (BZ) and cyclohexane (CH) at 50 °C for different times (a) and for PP fibers exposed in C_4 distillation condition (b).

The DSC results for the PP fibers exposed to various solvents are summarized in Table 1. The reported T_g for PP is -3 °C[10], however for all of the samples, there was no discernable glass transition observed between -30 and 40 °C. The lack of the observed T_g could be due to the high crystallinity, which tends to decrease the magnitude of the heat flow change at T_g, as well as to broaden the T_g so that the transition is indistinguishable from the baseline. For all samples, the only observed transition was a melting endotherm between 120 and 180 °C. While the peak in the melting endotherm did not change significantly with either solvent or aging, there were changes in the level of crystallinity. For both solvents the level of crystallinity initially increases and then after 12 months remains relatively constant.

Table 1. Summary of DSC results for as-received and treated PP membranes.

Sample History	**T_m (°C)**	**ΔH_f (J/g)**	**% Cryst**
As received	160	90.3	43
12 month in C_4 under distillation condition	161	103	50
Exposure to cyclohexane at different times at 50 °C			
1 month	161	69.5	34
3 month	161	87.9	43
6 month	161	96.6	47
12 month	161	106	51
18 month	161	101	49
24 month	161	105	51
Exposure to benzene at different times at 50 °C			
3mo Benzene 50C	161	95.8	46
6 month	161	73	45
12 month	162	105	51
18month	161	102	49
24 month	161	103	50

Dynamic mechanical analysis (DMA) was employed to examine the effect of solvent exposure on the mechanical properties of the PP hollow fibers. If solvent exposure causes a change in the polymer's crystalline structure, the mechanical properties of the membrane should reflect these changes. For each sample, an oscillatory strain was applied to the membrane in tensile mode and the resultant stress was measured. The oscillatory nature of the test allows us to probe both the elastic and viscous components of the material and obtain the complex modulus, $E^* = E' + iE''$, where E' is the storage modulus (elastic component) and E'' is the loss modulus (viscous component).

In Figure 4(a), the DMA results of PP fibers are presented after they were exposed to liquid C_4 under the distillation conditions. Oscillatory strain sweeps beginning at 0.01% strain and continuing until breakage were applied to each sample at a frequency of 1 Hz. In Figure 4(a), the slope of the line at low strains represents the Young's modulus. In this case, the modulus is relatively unaffected by the exposure time. There is a turning point at which the slope change marks the transition from a linear viscoelastic response (to the applied strain) to a non-linear response. The results suggest that the slope after this transition slightly increases after 12 months exposure. The storage (elastic) and loss (viscous) components of the modulus are shown in Figure 4(b). The results suggest that the soaked PP not only gives a slightly higher strain, but also gives higher storage and loss modulus.

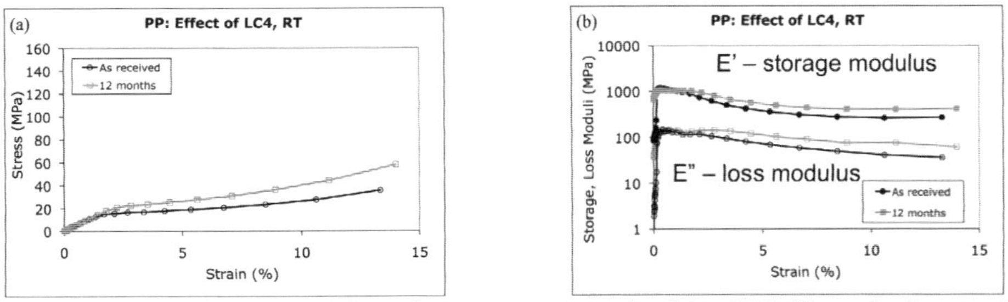

Figure 4. The oscillatory strain sweep (a) and storage *vs.* strain (b) of PP fibers before and after they were exposed to liquid C_4 for 12 months.

The results of the cyclohexane exposure are shown in Figure 5 where the slope of these lines is relatively unaffected by the exposure time. The transition point stays more or less the same. However, the strain at the breaking point does change with the exposure time. In the first 12 months, the breaking strain gradually increases as the exposure time increases. After the strain reaches its largest values for this increase, ~ 17%, it starts to decrease slightly after 24 month exposure. Overall, the behavior of stress *vs.* strain of PP fibers does change.

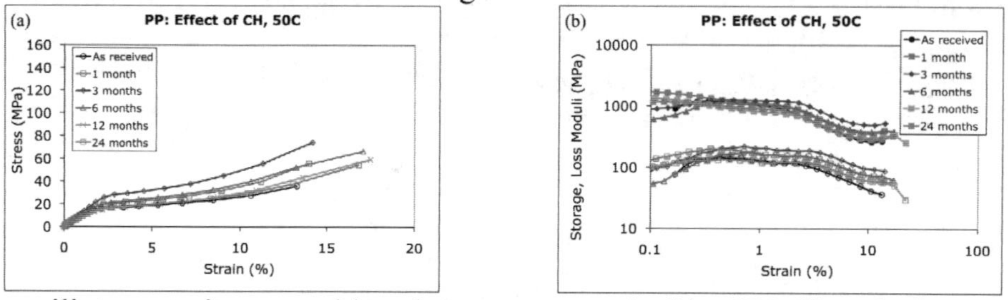

Figure 5. The oscillatory strain sweep (a) and storage *vs.* strain (b) of PP fibers before and after they were exposed to cyclohexane at 50 °C for different times.

From both SEM and BET results (not shown), one notices some rearrangement of the material's pore structure: the pores become smaller from 0 – 3 months and then begin to shift back to larger sizes in the meso-pore range. Correspondingly, the correlation of storage (loss) modulus *vs.* strain, as shown in Figure 5(b), also changes. However, comparing the 24-month exposure data, we find that the DMA data is almost identical to that of the 12-month exposure sample. The rearrangement of the pore morphology and crystallinity may be responsible for the changes seen in the stress-strain data.

In contrast to the aliphatic hydrocarbon fluids that show only induce modest changes to the fiber's mechanical properties, exposure to benzene significantly increases both stress *vs.* strain behavior and modulus, as shown in Figure 6(a) and 6(b), respectively. SEM (not shown here) shows that the pore morphology changes with increasing exposure and the membrane appears to be dense. Since the DSC results show similar crystallinities for the fiber samples exposed to both the aliphatic and benzene solvents, the increase in the tensile properties is most likely due to the changes in the morphology of the membrane.

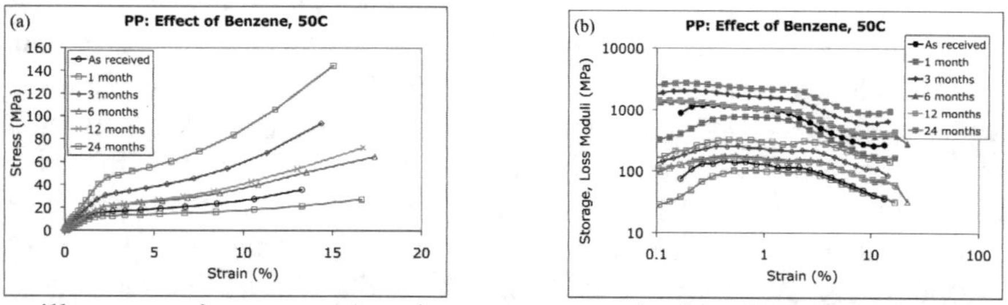

Figure 6. The oscillatory strain sweep (a) and storage *vs.* strain (b) of PP fibers before and after they were exposed to benzene for different time at 50 °C.

From the above study, we conclude that PP fibers are stable at iso-/n-butane environment for a long period of time. Therefore, PP fibers are suitable for the structured packings in ethane/ethylene, propane/propylene, and iso-butane/n-butane distillations, but not for the aromatic environments.

Conclusion

The preliminary results on two lab-scaled systems show great improvements in the separation efficiency and the apparatus capacity using hollow fibers as structured packings over the conventional packings in the olefin/paraffin distillation. In this scaled-up system, we demonstrate that the high separation efficiency (HTU < 30 cm) and a long-term stable operation (> 9 months) are achieved using hollow fiber as structured packings for the iso-/n-butane distillation. Up to 35% enrichment of iso-butane is achieved in a hollow fiber column with a 90 cm length. When the F-factor is up to 2.2 $Pa^{0.5}$, the pressure drop on the vapor side is below 20 mbar/m. This important result confirms that the pressure drop is not an engineering barrier for utilization of hollow fibers as structured packings in distillation columns. Furthermore, we also conducted the distillation experiments under changeable reflux conditions (reflux ratio > 8). The results suggest the experiments are not only more stable, but also give better separation compared to those obtained under the total reflux condition. We also found that the packing density is an important parameter in the module design for the stability and separation efficiency of the hollow fiber column.

However, as the system is further scaled up, the length of the column will increase. For longer columns, the pressure distribution along the column length will be critical to the operational stability. In addition to the packing density, the morphology of hollow fibers will play an important role in the module design to maintain the uniform pressure distribution between two phases, and thus ensuring high separation efficiency and operational stability. We anticipate more engineering challenges as the system capacity is increased into a pilot plant scale. However, given the significant improvement in separation efficiency, the prospective gains of using this novel technology are still very economically and environmentally attractive.

Acknowledgement

We gratefully acknowledge the DOE office of Energy Efficiency and Renewable Energy – ITP program for funding this work. We would like to thank Professors Cussler and Sirkar for their intriguing discussions and scientific insights about this technology. Also, we would like to thank the supports from Chevron Company. Finally, we would like thank Dr. Malcolm Morrison for providing fruitful discussions on the hollow fiber module design and a marketing perspective on the hollow fiber industry.

References

1. Eldridge, R. B., Olefin/paraffin Separation Technology: A Review. *Ind. Eng. Chem Res.* **1993,** 32, 2208-2212.
2. Chapas, R.; Colwell, J. A. *Industrial Technologies Program Reserach Plan for Enerngy-Intensive Process Industries*; Battelle: 2007.
3. Angelini, P.; Armstrong, T.; Counce, R.; Griffith, W.; LKlasson, T.; Muralidharan, G.; Narula, C.; Sikka, V.; Closset, G.; Keller, G.; Watson, J. *Materials for Spearation Technologies: Energy and Emission Reduction Opportunites*; DOE, EERE Office: Washington, D. C., May 4, 2005; p 103.
4. Kister, H. Z., *Distillation - Design*. McGraw Hill: New York, 1992.
5. Yang, D.; Barbero, R. S.; Devlin, D. J.; Cussler, E. L.; Colling, C. W.; Carrera, M. E., Hollow fibers as structured packing for olefin/paraffin separations. *J. Membr. Sci.* **2006,** 279, (1-2), 61-69.
6. Sirkar, K. K., Membranes, phase interfaces, and separations: Novel techniques and membranes-an overview. *Industrial and Engineering Chemistry Research* **2008,** 47, (15), 5250-5266.
7. Yang, D.; Devlin, D. J.; Barbero, R. S., Effect of hollow fiber morphology and compatibility on propane/propylene separation. *J. Membr. Sci.* **2007,** 304, (1-2), 88-101.
8. Yang, D.; Matinez, R.; Fayyaz-Najafi, B.; Wright, R., Light hydrocarbone distillation using hollow fibers as structured packings *J. of Membrane Science* **2010,** 362, 11.
9. Olujic, Z.; Behrens, M.; Sum, L.; Fakhri; Graauw, J. d., Augmenting distillation by using membrane based vapor-liquid contactors: An engineering view from Delft. *J. of Membrane Science* **2010,** 13.
10. Wunderlich, B., *Thermal Analysis of Polymer Materials*. Springer-Verlal: Berlin, Heidelburg, 2005.

Abstract

Jim Fair: Distillation – High Expectations

By

Daniel E Steinmeyer

Chem Engineering Energy

Chesterfield, Missouri

Presented at the James Fair Heritage Distillation Symposium
AIChE Spring Meeting, Chicago, March 2011

Jim Fair: Distillation – High Expectations

Dan Steinmeyer 1-22-11

"Every engineer should have a technical hobby or two..."
Jim Fair circa 1975

Jim Fair was a man of large dimensions, including tall height, long life, huge publication count and wide smile.

Jim won at least seven named awards from the Chemical Engineering community in a career that spanned 70 years, with about half spent at the University of Texas and half in industry, primarily with Monsanto. Jim's distillation course was taught 125 times over a 25 year period. His technical contributions were recognized by election to the National Academy of Engineering

But his largest dimension is with those he influenced throughout his life -- those who worked for him, as well as his students.

To borrow a quote about Jim from a memoir I wrote:

> *"A great boss even though I never worked directly for him ... many pieces of excellent advice ... responsible for more of the positive things that happened in my story than anyone else ... intelligence ... personal charm, perspective and ability to use words well."*

Other notes -- from others who worked with or for him:

> *"One of the pillars of our chemical engineering profession ... very keen, meticulous and showed a remarkable level of interest in people ... made it a point to come up and talk with me personally ... affable presence in the many things we enjoyed doing and learning from him."*

> *"He was my mentor and my friend. He, more than anyone else, shaped my career. The chemical engineering profession has lost one of the great ones".*

> *"... a great man ... put together the best and brightest engineering technology organization in the chemical industry ... things that I learned from him, particularly about organizations and people management, that carried with me throughout my career. He was blessed to have spent the last 30+ years of his life doing what he really loved - advancing technology and teaching people."*

> *"He taught by example and led us to do great work: do it right technically; take care of the details; prepare well written documentation."*

> *"He was a great mentor ... door was always open ... would stay late to talk ... helped me find my career path."*

He was one of those people whom others wanted to be around because it just felt good to be in his presence.

Distillation -- High Expectations

Jim encouraged us to ***be the best***. He also encouraged "technical hobbies". Distillation was his primary one although he also meandered into heat transfer, since like it or not, it can't be avoided in distillation.

If Jim could speak to "distillation" today, *as if* it lived and breathed, he would give it some of his "good perspective". He *might* tell it that it had been richly endowed by Mother Nature, was ***the best*** separation technique and should raise its ***expectations***. Jim, the son of a Presbyterian minister, was an exceptionally positive leader -- and ***sought the best*** from ***everyone and everything***.

He might detail the rich ***endowment of distillation*** and maybe sketch a picture:

- large density difference between phases and high interfacial tension -- which make it easy to obtain efficient multistage processing
- relatively high diffusivities – which enables high stage efficiency
- utilization of thermal driving forces hidden behind the flowing streams
- 100 year old long learning curve
- fully calculable – except for ….

Distillation Hobbies That Jim Bequeathed

Jim was directly responsible for most of my "technical hobbies". He would point in a general direction and say

"… you know … why don't you … there was a paper by … did you ever look into …"

or when he was being more directive

"… we need …"

This paper speaks to a number of these hobbies. They became *my* hobbies but they all trace to an input by Jim. They started with design details but grew up into bigger things like ***thermodynamic efficiency*** and how it is easily lost from view by low ***expectations***. This in turn stretched to include ***economic efficiency*** and optimized use of capital. Both efficiencies look outside the big vessel to the support systems for distillation, like heat exchangers and wander into the utilities that move heat in and out – and which, at optimum, usually dominate both capital and energy.

Along with his skills and talents **Jim** was tolerant. In my case his tolerance enabled me to pursue these hobbies until I almost got the story right. The ***hobbies*** that Jim led me into:

- Entrainment and generation of droplets in gas-liquid systems
- Mass transfer – what limits and why?
- Second law efficiency – and what it says about distillation systems.
- Economic efficiency and what it says about big columns and small reboilers.

Full-Scale Plant Column Efficiency - Measurement and Management

ID# 212830
Sean Hennigan -Separations and Heat Transfer Advisor , BP Chemicals, UK

Abstract

Examples of measurement of full scale plant efficiency (from process data, modelling and benchmarking) are offered from three different petrochemical areas. For each, an efficiency deficit was identified and linked back to a shortcoming in tray design and/or operation. In all cases this lack of efficiency is hidden from the operation teams who have often lived for years with these deficits unknowingly. This paper offers methods to tease out efficiency numbers, offers potential explanations for each and some practical rules to manage this uncertainty and the impact on the broader cost of production.

1. Overview

Tray efficiency must be the most elusive concept in distillation and ironically is the third key pillar of good column design along with the Vapour-Liquid Equilibrium and Hydraulic models. Even today, fresh and enthusiastic process engineers never fail to be bemused at the idea that the first two are converged with accuracies to the third decimal place, only to be joined by a third idea that is at best a limited model and at worst a guess.

Many excellent texts and models have been published on the subject, but this paper will not attempt to offer any more insight into the detailed theory. The intention here is to highlight practical methods of assessing shortfalls in efficiency through three real examples where methods are applied to attempt to measure, or at least compare, this tricky concept. Through the paper several rules of thumb are offered to help manage the uncertainties of efficiency.

At the simplest, overall tray efficiency can summarized by the following equation for *Overall Efficiency.....*

$$Eo = N_{theoretical} / N_{actual} \qquad \text{Equation 1}$$

Life is never perfect so, unless they are incredibly lucky, the designer will almost always need more actual trays than theoretical stages hence $Eo < 1$.

Eo is built up from three elements as follows...

- Point efficiency E_{OG}- how close to equilibrium a small chunk of liquid and vapour will become as they collide together on the tray deck.
- Tray efficiency (usually Murphree) E_{MV}- how this efficiency is enhanced on the large scale through the effects of flows and volatility on the driving forces.
- Overall efficiency *Eo* - how this is then further enhanced, or otherwise, by looking at the bulk effect on all stages and accounting for hydraulic issues such as entrainment or weeping.

Rigorous analysis at each point is challenging, and usually limited to binary systems but the author has found clear explanations in several excellent texts [i] [ii]

A rate-based approach is also possible that looks at tray performance from a fundamental mass transfer approach and is arguably far more rigorous, but often requires a great deal more detailed information and model calibration.

For an industrial practitioner, Equation 1 is a simple measure of the headline performance that is needed, regardless of the fundamentals that build it up. Of course all trays in a tower need not have the same efficiency, nor be the same for all components and for multicomponent systems things get very complex, all of which acts a barrier to getting a practical view. The author has found that returning back to Equation 1 is the best basis for rapid assessment of performance, despite its simplicity, and has successfully revealed otherwise hidden column problems.

For existing units, knowing if your tower is performing with a lower or higher than expected efficiency is valuable, especially in a world where energy and downtime is expensive. This can also offer capex reductions on future projects if overdesign can be removed. For these units we also have the advantage of available data and hopefully a process simulator, such as ASPEN Plus ™ to match to plant performance.

2. Overall Efficiency Assessment from Correlation and Theory

The first question is really around "expected" efficiency for a system. This implies an understanding of a theoretical efficiency to target and the author tends to use the following simple approach from O'Connell to provide at least a starting point for expected efficiency [iii]

$$Eo = 51 - 32.5\log(\mu_a \alpha_a) \qquad \text{Equation 2}$$

This broad correlation provides an estimate of overall efficiency based only on the components' average viscosity and relative volatility, implying that detailed tray design offers only marginal benefits. Closer examination shows this approach to be logical; more viscous liquids will suffer a reduced liquid diffusivity and hence a more controlling liquid mass transfer resistance. High relative volatility raises the equilibrium position and places more emphasis on the liquid film resistance in the overall number. This reduces the tray effectiveness but requires less theoretical stages to achieve a defined split. The latter idea is often counter intuitive, but makes sense when one remembers that "low efficiency" and "hard to separate" are really two different concepts.

Once the fundamental mass transfer is addressed any further enhancement of efficiency is really down to achieving plug flow on the deck. As shown by several authors [iv] large improvements are only really possible for systems that already have a high point efficiency, in excess of say 60% .

In the author's industrial experience the bracketed group in equation 2 above tends to be in the range 40-70% due to use of liquids with fairly water-like viscosity and reasonably low relative volatilities based on key components. This also bears out with most measured efficiencies of good working columns in the field (by the *process simulator method*, to be discussed) and most process simulators used for these systems are based on an overall efficiency in this range.

These assertions above have been further tested against theory in the following analysis.

A procedure for assessing the likely overall efficiencies of typical trays has been developed from a method defined by Lockett, Zuiderweg and others ([v]). From hereon denoted the "*Lockett method*". This is quite detailed and is only briefly mentioned here for reference as a purely theoretical cross check of more practical methods described later.

This method assumes trays of a "Lewis Case 1" type with good liquid and vapour distribution; flooding and weeping effects were assumed to be minimal, hence the results show what is possible from an ***adequate*** tray design. Several hundred data sets of physical properties, geometric and loading factors were generated in excel using a "Monte-Carlo" type approach to gain a random picture of likely parameters to be fed into the efficiency calculation. Each variable was bounded at high and low values based on the author's experience and a random value chosen for each run of the calculation. The Lockett method requires an idea of the "number of mixing pools" (*Np*) , derived from the *Peclet number* (*Pe*) to describe the approach to plug flow on the tray - this was left as a floating input.

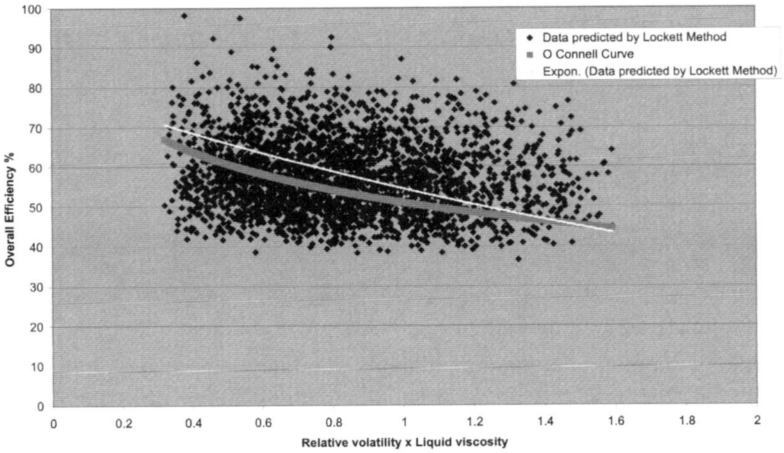

Figure 1 - Comparison of the Lockett method with the O'Connell Correlation

Typically for the cases considered *Pe* averages around 200 ,so this was used as the basis for the data in Fig 1 above.

Scatter is unsurprisingly huge but the curve is centred around O'Connell's equation and is bounded by the 40-70% rule of thumb discussed earlier. Increasing *Np* from the normal range expected to a situation approaching theoretical pure plug flow (*Pe* > 1000) shows *unnoticeable* improvement, again unsurprising as *Pe* in the hundreds implied near plug flow conditions anyway. This is shown in the figure below where the various best fit lines to the Lockett method have been plotted for varying *Np*. The only real improvement is if *Np* increases from 1 to about 3-5, ie from complete backmixing to a "hint" of some plug flow. Even then the size of improvement is within any reasonable measurement error.

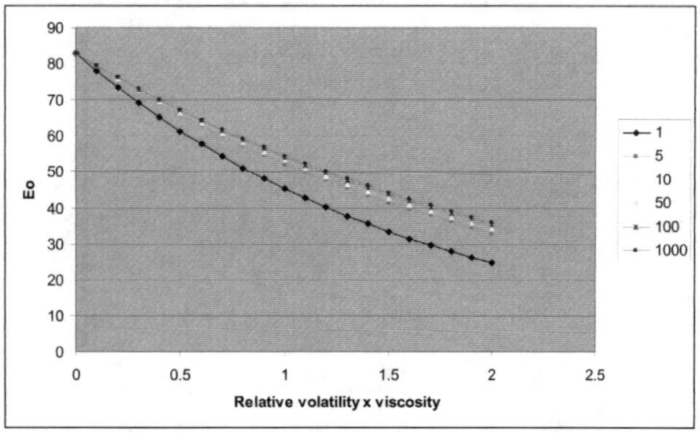

Fig 2- effect of plug flow on theoretical efficiency

The key point from this is that in both theory and practice, only *very small* improvements in efficiency can be obtained from trays ***working within hydraulic limits***.

Rule #1 - for a new system expect and design for 40-70% for a typical efficiency for a system around the middle of the O'Connell curve.
Rule #2 - Don't hope to get much more efficiency by modifying a hydraulically sound tray design.

3. "Process Simulator Method" for assessing efficiency

A view of the column overall efficiency from plant data can be easily achieved if a rigorous process simulation is available. The author has formalized the following method from existing less formal practices and found it applicable on many occasions.

1. Set up a standalone model of the column in question using the best thermodynamics available and an initial guess of the number of theoretical stages.
2. Choose a key composition that needs to be controlled (eg. lights in the base, heavies overhead etc) and set a design specification in the model such that the main energy driver (reflux ratio, reflux rate, boilup ratio or reboiler duty) is varied to always maintain this composition.
3. Step through a different range of theoretical stages in the column and plot the resulting energy driver against the ratio of these theoretical stages over actual trays in the field- this is Eo. This is represented by the blue curve below.
4. On the same graph construct a horizontal straight line corresponding to the plant-measured energy driver (red line of reflux rate in the example below).

5. The overall efficiency is where the curve crosses the line.
6. Any known error on the measured energy driver can be represented as error bars or a band whose width matches this error. This will then clearly show the range of error on the measured efficiency.
7. This method can also be worked by fixing energy input and matching compositions, but relies on the flows being very accurate.

If the column is operating close to minimum reflux, then getting a good handle on efficiency can be difficult as the energy driver curve flattens off as reflux falls. High column efficiencies (> 50%) can therefore be difficult to quantify, conversely if the column has a very low efficiency (< 30%) this will be quite obvious as the reflux will be driven very high near a minimum stages condition.

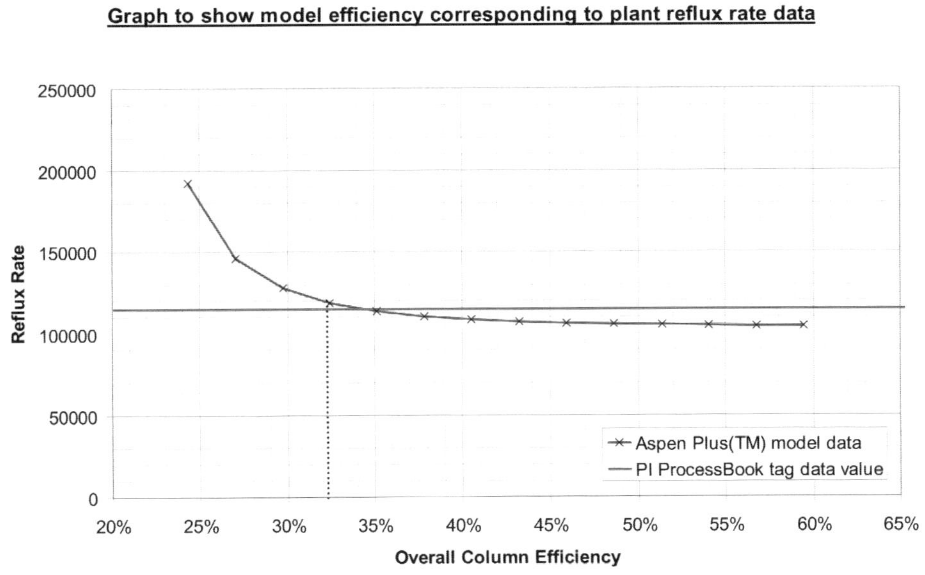

The method can also be used to look at different efficiencies above and below the feed and with further tweaking, efficiency in different column sections is possible to try to match model temperature profiles to tray temperature data. The latter are often thin on the ground and sometimes confused by pressure compensation, so care is needed if the investigator digs this far down.

The author has found changing the theoretical stage count a more robust match to plant data rather than the option of setting a stage-efficiency within the simulation, which often gives strange temperature profiles.

This approach has generally yielded the 40-60% *Eo* as predicted by O'Connell, which for the high *Pe* systems we have suggests E_{OG} are a similar magnitude, which on several occasions has been matched by the Lockett method. ***The implication from this is therefore any measured efficiencies falling well below this curve are an indication of a major tray hydraulic problem.***

3.1 - Process Simulation confirms low efficiency

The simple method above has been applied many times by the author to real examples. The case described below is a real example and typical of the method and here the tower sits in the middle of a very complex flowsheet and separates a whole range of components, making it very difficult to apply any kind of shortcut method for efficiency. In this case the author has found that the method above is the only realistic option.

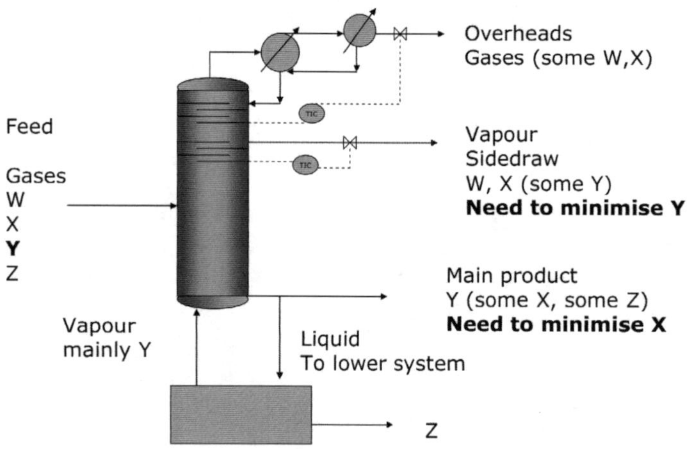

Figure 4 - Example column layout for the process simulator method

The main concern with this column is the slippage of useful lights to the main product base which can end up recycling round the main process causing bottlenecks. Also the vapour side draw can see high levels of heavies which reduces the value of this stream in further processing.

The efficiency was determined by the process simulator method, but the unusual set up of design specs required a match the slippage of X and Y into the light sidedraw and the base respectively, rather than a match to reflux.

The results showed that the base section efficiency (below feed) had an *unnoticeable* effect, implying that the key separation is occurring in the heads. The top section split of X and Y showed a large dependency on efficiencies in this section

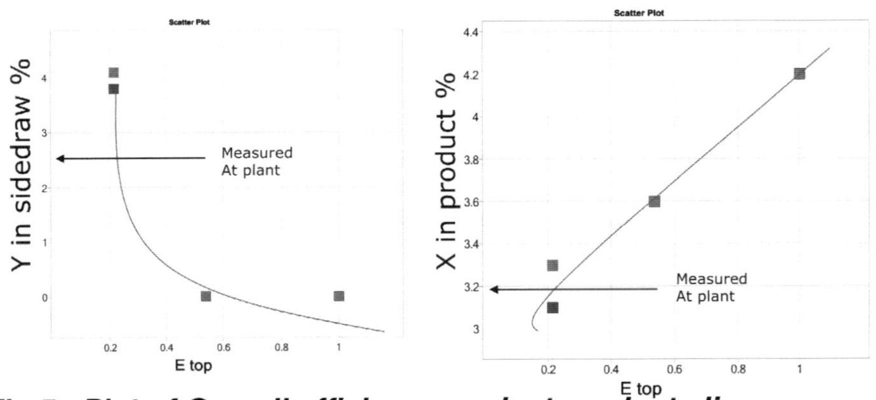

Fig 5 - Plot of Overall efficiency against product slippage

Figure 5 above shows the simulated effect of changing the top section efficiency (E top) on the levels of X and Y in the product and sidedraw respectively. The different colours represent different base section efficiencies which has negligible effect on the overall separation. The best match to plant data occurs at a head efficiency of around 20-30% and other checks were also carried out to match temperature profiles and vapour sidedraw flow all giving similar efficiency levels. The actual slippage of Y is much higher than desired which actually increases the sidedraw flow and forces a lower than expected loss of X due to this "cut" effect. Looking at the slippage of X in isolation could lead to a false sense of security on efficiency.

For every tray, a theoretical point efficiency was also calculated from the Lockett method from physical properties, flows and geometry data. This was converted to a worst-case overall efficiency assuming that the trays are fully backmixed (ie no plug flow benefit). All of these numbers are in line with the O'Connell prediction. The results in fig 6 below show a large discrepancy between theory and overall measured efficiency, even much worse than theoretical point efficiency (base measured efficiency was assumed to be 30% - although cannot be fully concluded).

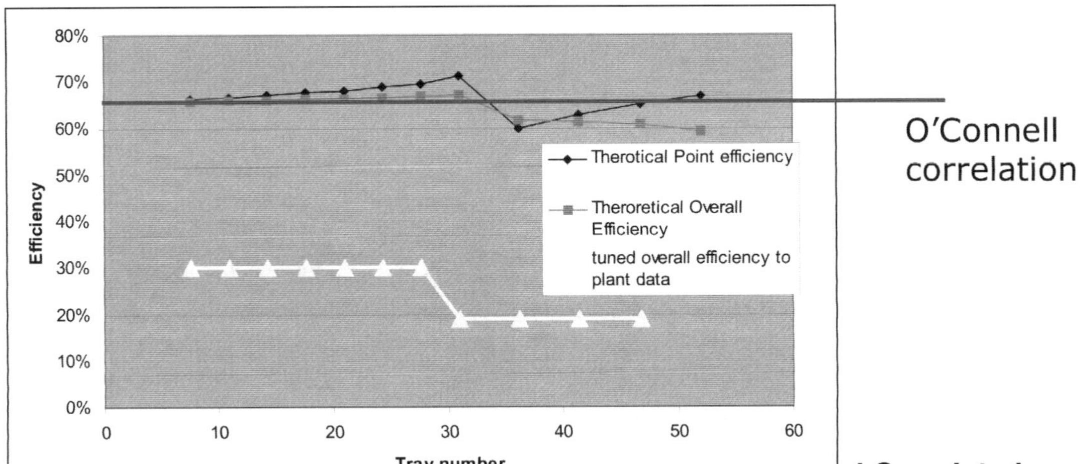

d Correlated

3.2 <u>Root cause of low efficiency</u>

Analysis of the design tray data sheet showed two main problems. This tower had been optimised over several years in a very successful effort to reduce energy usage that resulted in much reduced internal load in the column. The column is equipped with a set of 2-pass sieve trays at the top with a hole area in excess of 14%, plus 3-pass trays for the lower section with a hole area around 13%.

A detailed view of the tray design showed the potential shortcomings of these trays. Internal loads and properties were derived from the process simulator and hydraulics checked using the FRI Device Rating programme and vendor software. Particular issues were noted for the top section and summarized below

% Jet Flood	**52%**	
Weep fraction	**46%**	
Approach to weep point	**145%**	**(less than 100% → no weeping)**
Dry drop/wet drop	**0.28**	

The trays are weeping significantly. A closer look at the % hole area shows that these trays were designed many years ago with little turndown in mind. It is likely that these trays are also suffering maldistribution as the ratio of dry pressure drop to clear liquid height is very low.

These 3 pass trays at the column base were predicted to be highly loaded from the vapour side..

% Jet Flood	**86%**
Weep fraction	**2%**
Approach to weep point	**130% (less than 100% → no weeping)**
Dry drop/wet drop	**0.3**

These trays are close to jet flood, but also are also simultaneously on the theoretical weep point, but little volume of weep predicted. Premature flooding is a possibility in this situation. The dry/wet drop ratio is also very small and add to this the fact that the 3-pass trays have no flow balancing to get even L/V ratios on each panel and there is serious potential for efficiency shortfall. The predicted Bolles ratio for this system is in excess of 1.4.

The technology had been pushed hard to save energy on the system and was successful in delivering this, however as we chase down the more subtle savings it became clear that a legacy of suboptimum separation performance existed as the trays had drifted into weeping. Keeping component X out of the main product flow avoids recycle and energy costs. Keeping Y out of the vapour sidedraw increases the value of this export stream.

The current low efficiency prevents the plant from doing both at the same time, but the symptoms of this were not obvious, particularly against a background of changing feedstock quality which produced swings in the level of X in the product. Returning the efficiency back to a reasonable number offers significant value and rapid payback on the recommended retray. 3 and 2 pass trays will be retained to minimize capital but the design will be changed to a balanced movable valve tray eliminate weeping and ensure equal L/V on each panel. At the moment this change is planned for a future turnaround opportunity.

Rule #3 - Check efficiency <u>in depth</u>, particularly of older kit - they may reveal a performance deficit that has been lived with for years.

4. "Shortcut Fenske" method to compare "before and after" efficiency

An alternate method based on varying approximations of the Fenske relationships can be employed that dispenses with the need for a simulator as it needs only plant data. This works best with systems that are pseudo-binary and with compositions that are frequently measured.

In its simplest form put this suggests the following relationship for binary separation ([vi])

$$\ln S = f_n\{R, q, N\} \qquad \text{Equation 3}$$

$$S = (x_{LK} / x_{HK})_D (x_{HK} / x_{LK})_B \qquad \text{Equation 4}$$

S = separation factor
x = mole fraction
R = reflux ratio
N = number of stages
q = column feed state
Subscripts
HK = heavy key
LK= light key
D = distillate
B = base

The function on the right hand side of equation 3 can vary in complexity, but ideally should be a measure of energy input to the system. Generally as *N* or *R* increases with *q* unchanged one would logically expect *S* to increase. Using ln*S* as a measure of performance is incredibly useful and often overlooked. Engineers and operators are very often concerned with only one stream, subtle changes in the other stream may suggest a big shift in efficiency that is nearly invisible, it is important that the overall effect on both *D* and *B* are understood.

For small changes in reflux ratio (as usually seen on operating plants) the relationship between ln*S* and *R* seems near linear, if this behaviour shifts after a shutdown or upset then it implies that *N* and hence efficiency has changed.

4.1 Shortcut Fenske Example

In this example the tower was regularly sampled at both the top and both products and acted broadly as a near binary separation. This was modeled as a binary system in the process simulator, but also actual plant performance was fitted to a simple equation based on equations 3 and 4 .

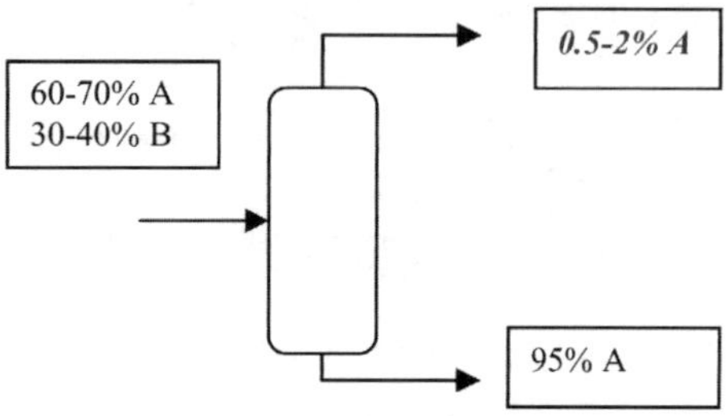

Fig 7 - Example Tower for Fenske Shortcut analysis

The operator's focus is on minimizing the slip of A over the top and concerns were raised after a debottleneck when conditions seemed to be different. The tower had been revamped with sections of higher capacity trays, upon restart the full debottleneck rate was not achieved (for several reasons unrelated to the tower), but it was "felt" by the operators that there had been a shift in the tower's separation performance. At the same time the feed composition to the tower had also changed slightly as part of the debottleneck, making it difficult to conclude any real problems.

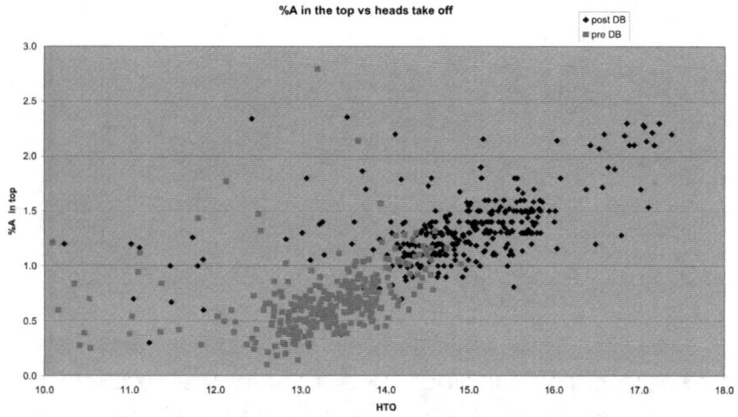

Fig 8 - Pre and post shutdown performance, slip of A vs heads flow

Post debottleneck the column heads take off was higher, but so was the level of A in the feed. One could expect more heads take off if the level of A in the base remained high. At face value the curve above seemed to imply tower performance was unchanged and was simply reacting to a different feed condition. This is only half the story as there is no measure of energy input to the system, or any analysis of changes to the tower base.

A much deeper insight was obtained from the method of equations 3 and 4. The separation factor *S* was calculated from both top and base compositions and plotted against reflux ratio.

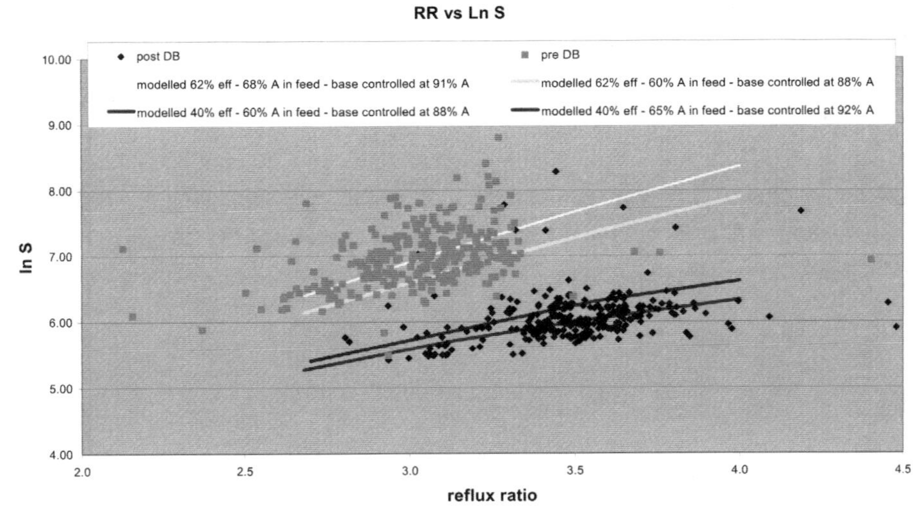

Fig 9 - Fenske Analysis for the same column data

Clearly the pre- and post- debottleneck data is very different; now more reflux ratio was needed to achieve the same separation factor. To be sure that the method was accurate, it was confirmed against the process simulation method (for both pre and post debottleneck conditions) which was simplified to reflect the near binary separation.

The simulation method result is plotted as lines on figure 9 above and confirms a shift down in efficiency, even with a varying feed composition and allowing some uncertainty in the measured base composition. The simulations were run with a design spec on the base of 91 and 88% A and a 10% variation in the level of A in the feed, which covered the range of observed variation from the plant data. Even accounting for this there is an apparent efficiency loss of over 20 percentage points.

4.2 Root cause of efficiency change

The explanation was not totally clear but there were several candidates. The data above proved very useful in bringing the operator and the vendor together in a positive way to address the problem.

- The tray is single pass with a long flow path length. The capacity of the system was maximized by changing some highly loaded trays for a low pressure drop fixed valve design. There was some concern that this may have affected vapour distribution on the tray and the vendor suggested options, but any major work in the column was economically very unattractive.
- There was also some concern that the installed weir heights on some of the trays did not match the drawings and/or the design data sheet. At one point the drawing, data sheet and measured weir heights were all different, although in theory (by the Lockett method) this difference in weir height would not have produced the shift measured in figure 9 above (more like 5-10%). However it was decided to fix the weir height issue as a minimum effort to at least directionally help the efficiency.

- Shutdowns always make or break theories and it was found that the trays were all intact, weir heights were correct as designed, but inspection of the downcomer on a lower tray revealed a major surprise !

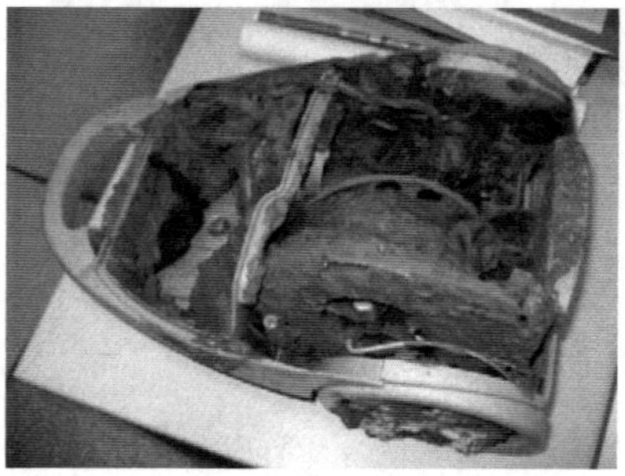

Fig 10 - A surprise finding in a downcomer

Not one but ***two*** vacuum cleaners had been left in the downcomer and remained boxed-in for two years. The plant is still restarting at the time of writing so it is not clear if removal of these has solved the problem. This tray near the bottom of the column close to where the composition is changing rapidly so it is possible that their presence affected the tower performance, but it also could be some influence of the low tray pressure drop, which may reduce as the system loads up to debottleneck rates.

As a postscript to this incident, the plant operations staff sensibly questioned the overall consequence of this low efficiency. On a stand alone unit more slippage of A overhead would represent a loss of valuable product and a cost, however on this plant the overhead stream went to an analogous column feed where there was a "second chance" to recover A on a sister plant with negligible economic penalty. Further simulation showed only a marginal financial penalty in terms of energy usage, and figure 9 suggests only a 20% penalty on increased reflux ratio to return to the higher separation factor.

As a result of this it is unlikely that the plant will pursue any further hardware changes in the near future. The analysis however proved valuable for future developments of similar systems and is a clear example of using this conclusive method.

Rule #4 - Tower performance "feeling" wrong is probably a strong indication of an efficiency issue.
Rule #5 - Measuring efficiency consistently before and after changes and shutdowns enables sensible conversations between vendor and operator.
Rule #6 - Follow through the actual financial consequences of both low and high efficiency- the answer may not be what you think.

5 - The "Benchmarking Method" to assess efficiency deficits in similar units

A third way of identifying low efficiency issues occurs if one is fortunate enough to have data from several columns of similar design to allow a back to back comparison of efficiency.

Benchmarking is a very powerful tool and the example below managed to shed new light on a long-standing problem.

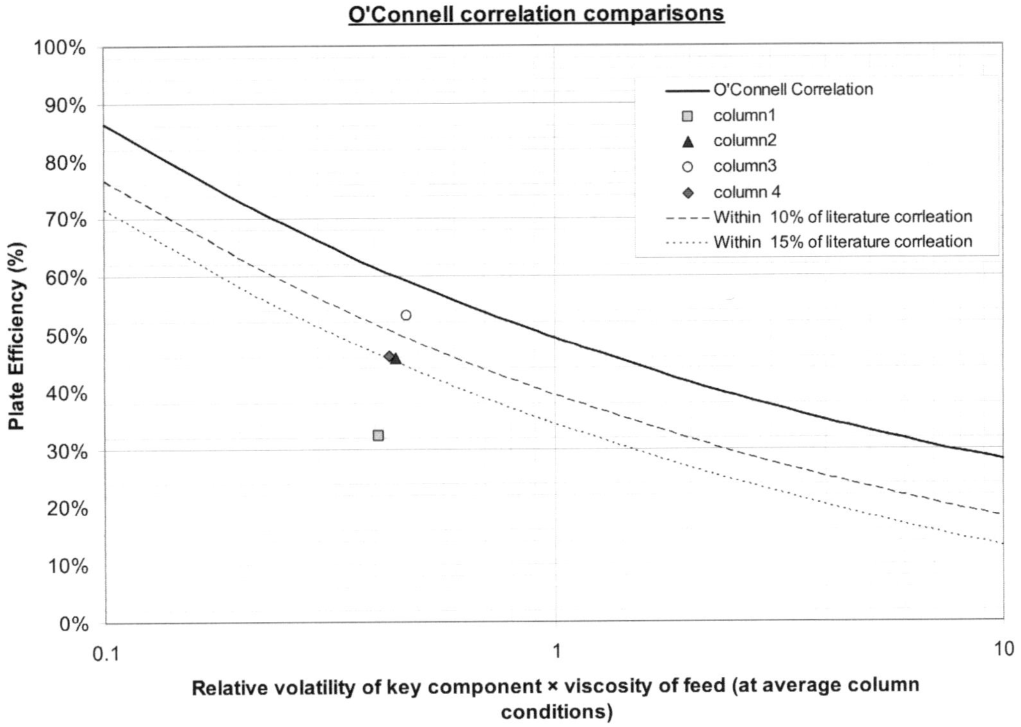

Fig 11 - Benchmarking against O'Connell

5.1 Example of Benchmarking

Columns 1 to 4 in the example above are in plants in different parts of the world but all essentially separate the same components with slightly differing feed and product composition, but all are pseudo-binary separations. For many years disparate groups worked on optimization of these columns with differing modelling approaches until a major study was undertaken to *benchmark* efficiency of these similar columns with a view to revealing best practice in operation and design. Column 1 in particular was felt by the operating plant that its performance could be better but was not well quantified, figure 11 above shows column1 to be a low outlier in terms of efficiency, certainly below the 40-70% rule of thumb discussed earlier.

Efficiency was calculated by the *process simulator method* from snapshots of steady running plant data. Other plots were used with increasing complexity (such as the Jafarey, Douglas, McEvoy equation) that all showed similar conclusions.

A remarkably simple picture emerged if the ratio of *steam usage per tonne product per theoretical stage* were used as a benchmark measure. In this example all the columns lie on a straight line if this value is plotted against $\ln S$. The error bars on the points represent 10% uncertainty and column 1 is the uppermost point - ie the highest energy input and the highest separation "output" . If the analysis of this column was flawed and the efficiency was more in line with the other columns then the performance would move to point "x", breaking this straight line relationship. This lends more weight to the assertion that the column is underperforming.

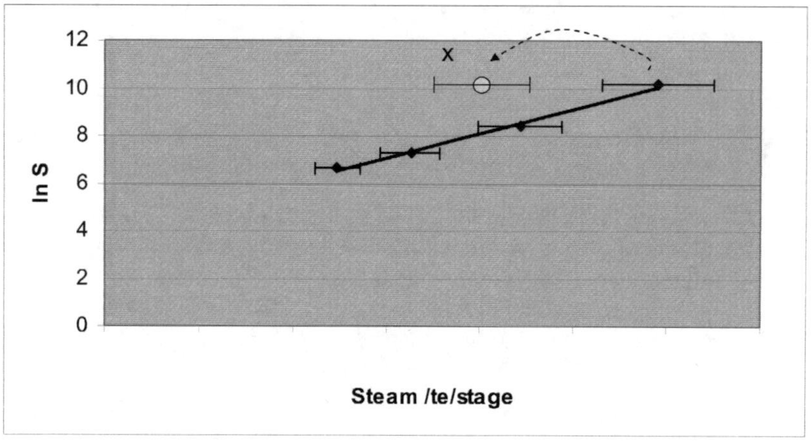

Fig 12 - Benchmarking lnS against Steam/te/stage

Column1 was due to be replaced with a larger unit to cope with future demand and the first pass approach was simply a direct scale of the current single pass design . This was strongly challenged once the results of this study came to light as there was a risk we could simply continue this low efficiency or possibly make it worse in the larger tower. The benchmarking tool proved a highly effective means of intervention.

5.2 Root cause of the efficiency difference

Column 1 was a wide, single pass column with sieve trays, the other units are smaller single pass or similar size and two pass, all are low pressure. Column 1 used sieve trays all the others were fixed valves. The problem column was theoretically a long way from flooding or weeping but stood out in having both a longer flow path length and higher weir load than any other unit.

These clues suggested a potential culprit in the phenomenon known as "Vapour Cross Flow Chanelling" ([vii]) - From hereon "VXFC". Under this condition the trays can effectively suffer simultaneous high entrainment and weeping due to unfavorable vapour/liquid distribution conditions. To be susceptible to VXFC a set of conditions need to be met..

- The column is operated at low pressure (< 5 bara)
- Exceeds a weir load 50 m3/hr/m
- Ratio of flow path length to tray spacing is greater than 2.5.
- The installed sieve hole area greater than 11%

Column 1 violated all of these limits with the exception of hole area which was installed at only 9%. However, detailed analysis of the column pressure drops revealed that the column DP had deteriorated steadily year on year and it was possible that the hole area may well have increased with time to a much higher number due to corrosion, a known issue with this column. Even without this mechanism the long flow path length on a single pass suggests some potential for liquid stagnation near the wall. ([viii])

Even today this root cause still remains a theory as there was no opportunity or funding to carry out a detailed column scan to prove this conclusively. However it was decided to design-out all potential for VXFC and liquid stagnation for the replacement column 1 and a two-pass fixed valve basis was used. This halved both the weir load and flow path length and kept the hole area within recommended levels. Column tray spacing, feed location and compositional specifications remained unchanged.

The new column was successfully commissioned and showed a significant reduction in steam usage as demonstrated in figure 13 below. Without the insight of the benchmarking data above we may have been in a situation with an even wider flow path length and possibly more susceptibility to VXFC and/or liquid maldistribution.

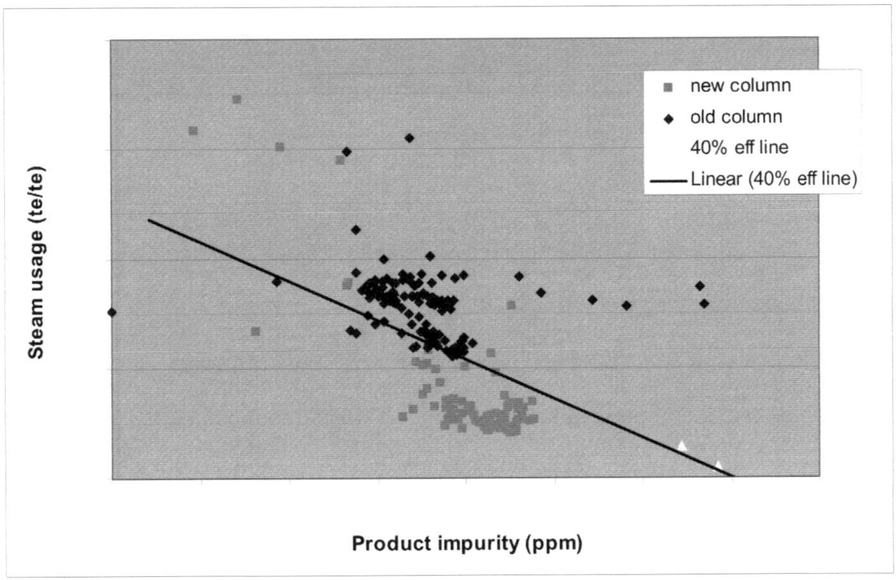

Fig 13 - reduced steam usage with bigger column, better tray design

Rule #7 - Look for similar efficiencies in families of similar separations. This is the most powerful way to detect deficits and paves the way for future changes.

6. Conclusions

Without very sophisticated models, tray efficiency will remain an elusive concept and a question mark in practical column design and operation. However there are methods and experience presented here that have been used by the author to compare and estimate full scale column efficiencies in the field with a view to identifying shortfalls and setting future designs.

These methods only require access to steady-state simulations of key areas of the plant, or even just manipulation of plant data.
Also from application of these methods, a broad set of rules of thumb for managing the efficiency question are offered (summarized in Appendix 1).

SA Hennigan

SA Hennigan January 2011

<u>**Appendix 1** - Rules of thumb for managing efficiency</u>

1. For a new system expect and design for 40-70% for a typical efficiency for a system around the middle of the O'Connell curve.
2. Don't hope to get much more efficiency by modifying a hydraulically sound tray design.
3. Check efficiency <u>in depth</u>, particularly of older kit - they may reveal a performance deficit that has been lived with for years.
4. If tower performance "feels" wrong this is probably a strong indication of an efficiency issue.
5. Measuring efficiency consistently before and after changes and shutdowns enables sensible conversations between vendor and operator.
6. Follow through the actual financial consequences of both low and high efficiency- the answer may not be what you think.
7. Look for similar efficiencies in families of similar separations. This is the most powerful way to detect deficits and paves the way for future changes.

<u>**Appendix 2 - Nomenclature**</u>

Eo = Overall Efficiency (%) or (-)
E_{OG} = Point Efficiency (%) or (-)
E_{MV}= Murphree Efficiency (%) or (-)
Pe = Peclet Number , ratio of forward to diffusive flow (-)
N_p = Number of mixing pools , derived from Pe (-)
S = separation factor, measurement of column performance (-)
x = mole fraction (-)
R = reflux ratio (-)
N = number of stages (-)
q = column feed state (heat to vaporize 1 mol of feed / molar latent heat) (-)
μ= viscosity (cp)
α= relative volatility (-)
L/V = liquid to vapour ratio (-)
VXFC = vapour cross flow chanelling
Subscripts
HK = heavy key
LK= light key
D = distillate
B = base

References

[i] Distillation Design (1992) , chapter 7- *H Kister*
[ii] Distillation Tray Fundamentals (1986) , chapters 8,9,10 - *M Lockett*
[iii] Chemical Engineering Volume 6 (1991) , page 442 - *JM Coulson, JF Richarson, RK Sinnott*
[iv] Distillation Tray Fundamentals (1986) , chapter 9 - *M Lockett*
[v] Distillation Tray Fundamentals (1986) , chapter 8, section 8.5 - *M Lockett*
[vi] Distillation Design (1992), chapter 3 page 106-126 - *H Kister*
[vii] Vapour cross Flow Channelling on Sieve trays Fact or Myth ? - Chemical Engineering Progress November (1992) page 87 - *Kister, Larson, Madsen*
[viii] Distillation Tray Fundamentals (1986) , page 183 - *M Lockett*

OVERALL EFFICIENCY EVALUATION OF COMMERCIAL DISTILATTION COLUMN WITH TWO-PASS VALVE TRAYS

P. L. Invernici

Basic Petrochemical unit, Braskem, Triunfo, RS

A. R. Secchi[*]

Chemical Engineering Program, COPPE, UFRJ, Rio de Janeiro, RJ

T. F. Mendes

Department of Chemical Engineering, UFRGS, Porto Alegre, RS

Abstract: This article addresses the prediction of the efficiency of two-pass crossflow valve trays in an industrial distillation column using different models with and without the use of the *Stagnant Regions Model*, as proposed by Lockett in 1986.Contradicting what could be expected, few papers have been published about the main characteristics of valve trays and about efficiency of two-pass valve trays. The results suggest that O'Connell's correlation, the model of Chan and Fair applied to two-pass valve trays converted to one-pass valve trays and the model of Chan and Fair with the *SRM* method are those that best predict the efficiency of this two-pass crossflow valve trays column. The O'Connell's correlation underestimates the efficiency, showing a slightly conservative value, while the predictions using Chan and Fair model and the modified Chan and Fair model with equivalent length proposed by Klemola and Ilme overestimate the efficiency, showing slightly non-conservative results.

Keywords: two pass crossflow valve tray, Tray Efficiency, Distillation Columns.

[*]**Correspondence concerning this article should be addressed to A. R. Secchi at arge@peq.coppe.ufrj.br**

1 Introduction

In simulation and design of distillation columns usually is employed the equilibrium stage model with the assumption that the vapor and liquid streams leaving the same tray are in equilibrium. The terms "ideal stage" and "theoretical tray" are both given to that stages. In real distillation columns, the conditions for vapor and liquid being in equilibrium are rarely achieved. The efficiencies are used to compare the separation of a real separation tray and that of a theoretical tray by introducing non-ideal factors into equilibrium stage separation models.

Tray efficiency is a crucial factor in the design and analysis of existing distillation columns.It relates the number of theoretical stages to the number of real trays. The normal procedure of designing a fractionating column is to calculate the number of theoretical stages with the help of rigorous models that are available in process simulation packages. The conversion to real trays may begin with a prediction of local efficiency (point efficiency), followed by transforming this value to the entire tray (Murphree efficiency), or by using the Murphree efficiency itself. The number of real trays required can then be calculated by converting the efficiency of the tray into the overall efficiency of the column.

Point efficiency is related to Murphree tray efficiency through the use of correlations that take into account the flow pattern of liquid over the tray. An important parameter for the design of a tray, which affects its efficiency, is the path length of the liquid[10].

It is a challenge to get the best sequence of methods and correlations to estimate the overall efficiency of a column, knowing that a unit operation as well studied as distillation suffers from a large amount of unsolved problems of efficiency prediction[7].

A one-pass crossflow valve tray to promote contact between vapor and liquid is the simplest type used in distillation and it is the type that is most commonly found in industry. However, when the liquid flow rate is proportionally higher when compared to vapor flow rate, or when the column diameter is large, it is necessary to use trays with two or three and sometimes even more passes to improve the hydraulic characteristics of the flow.

Few papers have been published about the main characteristics of valve trays[2] and about efficiency of two-pass valve trays[9,10,11,14,15].

The geometry of two-pass crossflow valve trays differs significantly from the geometry of one-pass trays. In two-pass trays, the liquid flows downward to the next tray through two side downcomers and then horizontally across two areas of vapor-liquid contact of the tray (two panels), pouring through the spillway into the central downcomer and so forth.

Several empirical and semi-empirical methods have been proposed for determining tray efficiency and point efficiency. In this study, special attention was given to the models of O'Connell[17], Chan and Fair[4], Garcia[6], and modifications in the latter two models by introducing the concept of equivalent length (C_E) proposed by Klemola and Ilme[11]. A more comprehensive review of works published on models for efficiency prediction can be found in the work of Invernici[8].

O'Connell's[17] empirical correlation has been used as an industry standard to calculate global efficiency of distillation columns for decades. It is based on liquid viscosity and relative volatility evaluated at average top and bottom temperature and feed column composition. The method predicts 90% of the efficiency data with a 10% margin of error for valve trays.

Designed for perforated and valve trays, Chan and Fair's[4] model is considered a refinement of the AIChE[1] model. Compared to the AIChE model, it was shown to be more suited for experimental data, with an average absolute deviation of 6.27% as opposed to 22.9% of the AIChE model. The model is recommended for use in simulators for being easy for convergence. It has an approach to the hydrodynamics of the tray that is less complex when compared to Garcia's[6] model.

Klemola and Ilme[11] proposed a method for calculating the Murphree efficiency of trays where the geometry of a two-pass tray is converted to the geometry of a one-pass tray, keeping constants the active area of the tray and the diameter of the column. The path length of the liquid flow of an equivalent one-pass tray should be considered as the sum of the path lengths of the liquid flows of a two-pass tray.

Garcia[6] presented a theoretical model to estimate the efficiency of perforated trays. It was a refinement of the model proposed by Prado[18], for non-aqueous systems, in which three regions were considered: the region close to the holes, where the gas may be in jet flow regime; the froth region, which is composed of gas bubbles dispersed in liquid; and the spray zone. Garcia and Fair[7] admitted that the model is more complex than the model of Chan and Fair[4] and more difficult to apply; however, the absolute deviation was 13.28 % as opposed to 18.22 % of the model of Chan and Fair[4].

In this study, we propose to evaluate the efficiency of two-pass crossflow valve trays by using several models that allow calculate the efficiency of even- and odd-numbered trays that have different geometries. It has the aim of expanding the studies of Domingues et al.[5], who sought to establish appropriate paths to estimate the efficiency of one-pass valve trays and dualflow trays.

2 Theoretic Fundamentals

The overall efficiency of a column is defined as the ratio between the number of theoretical stages required for a given separation and the number of real trays of the column at a given reflux ratio.

Point efficiency is defined by the variation of the molar fraction of a component on a point of the tray by the variation in the molar fraction of this component at the same point, which should occur, considering the theoretical stage.

The theoretical fundamentals of point efficiency have been presented by Lockett[15] and Kister[9]. The theory is based on the resistance to transferring mass between vapor and liquid films in a vapor-liquid interface. The point efficiency is obtained from the number of overall vapor-phase mass-transfer units (N_{OG}):

$$E_{OG} = 1 - \exp(-N_{OG}) \tag{2.1}$$

As the molar fraction of the component at a given point of a tray may not exceed the molar fraction of equilibrium, point efficiency (E_{OG}) is always less than 1.

The equation that defines Murphree tray efficiency (E_{MV}) is similar to the one that defines point efficiency, except that it takes the average concentration of the vapor mixture that comes out of and goes into any *n* tray.

$$E_{MV} = \left(\frac{\bar{y}_n - \bar{y}_{n-1}}{y_n^* - \bar{y}_{n-1}} \right) \tag{2.2}$$

Converting point efficiency (E_{OG}) into Murphree tray efficiency (E_{MV}) depends on using a liquid mixture model above tray n.

Generally speaking, in crossflow trays the value of Murphree tray efficiency (E_{MV}) is greater than that of point efficiency (E_{OG}). The liquid over the trays is never perfectly mixed in the horizontal direction, establishing a concentration gradient of liquid that moves across the tray. This, in turn, causes a gradient in the concentration of vapor leaving the tray. The existence of a gradient of concentration of a component in the liquid that flows over the tray results in an increase in the Murphree tray efficiency (E_{MV}), in some cases causing this efficiency to exceed 100%.

The patterns of a mixture of liquid over the tray could be plug flow, stagnant zones, or backmixing, where the drops of liquid projected by vapor change the concentration profile of the components in the horizontal direction of the flow over the tray, while the vapor flow patterns could be plug flow or perfect mixture[9].

Lewis[13] model, developed in 1946, is still the one that is most used for converting point efficiency (E_{OG}) into tray efficiency (E_{MV}). The work of Lewis[13], based on that of Krischbaum[12], who found considerable variations in concentration of liquid through the trays, was the first to develop quantitative relationships between E_{MV} and E_{OG} for situations in which there is mixing of liquid in the horizontal direction of the tray (plug flow), introducing changes in the method of McCabe-Thiele for rectifying binary mixtures[13].

The three cases considered by Lewis[13] are:

Case 1: vapor is completely mixed between trays. The direction of liquid flow in successive trays is immaterial.

Case 2: vapor is unmixed between trays. Liquid flows in the same directions on successive trays.

Case 3: vapor is unmixed between trays. Liquid flows in alternate directions on successive trays.

The correlations proposed by Lewis estimate maximum tray efficiency. In fact, the Murphree tray efficiency is less than that which was calculated due to deficiencies in the mixture of liquid on the tray, the non-uniform flow of liquid and vapor through the tray, and the effects of liquid entrainment and weeping.

From the Murphree tray efficiency one can estimate the column efficiency by using Equation 2.3, when the ratio of the slope of the equilibrium line (binary) to the slope of operating line is constant:

$$E_{OC} = \frac{\ln[1 + E_{MV}.(\lambda - 1)]}{\ln \lambda} \tag{2.3}$$

where λ is the ratio of the slope of the equilibrium line (m) to slope of the operating line (L'/G'):

$$\lambda = m.\left[\frac{\left(\frac{M_G}{MW_G}\right)}{\left(\frac{M_L}{MW_L}\right)}\right] = m.\frac{G'}{L'} \tag{2.4}$$

In binary mixtures, the mass transfer efficiency between two components is identical, but in multi-component separations the mass transfer efficiencies are different. Kister[20] detailed the reasons for these differences, pointing out the different diffusivities of the components, both in the liquid phase and also in the vapor phase, and for different slopes of the equilibrium curve and curves of operation of each component.

For multi-component systems, determining the slope of equilibrium can be done by using the procedure proposed by Chan and Fair[7]. The procedure, called a pseudo-binary method, is simple and easy to apply[20]. The only concern is that one must have some knowledge regarding the selection of light key and heavy key components[15].

The simplicity, good acceptability[9], and the ease of implementation have led us to adopt the pseudo-binary method in this work.

Since two-pass tray geometry differs from one-pass tray geometry, it is necessary to make special considerations for this type of tray.

Valve trays, with certain exceptions, behave in the same fashion as do sieve trays[2].

Bolles[2] proposed a method to predict the effect of tray geometry on the distribution of liquid and vapor among several passes and their impact on mean tray efficiency, as well as designing multi-pass sieve trays, valve trays, or bubblers trays with two, three, or four passes with optimal efficiency and flow distribution.

The results presented indicate that two-pass trays do not present special problems in distribution. Flows of liquid and vapor are distributed equally between the passes and average Murphree tray efficiency (E_{MV}) is essentially equal to the arithmetic average of the efficiencies of the individual passes.

However, the study of Bolles[2] does not address the reduced efficiency caused by the existence of stagnant regions of liquid on a multi-pass tray.

The most accepted model for assessing the effects of poor distribution of liquid on the plates for calculating tray efficiency was proposed by Lockett[15] and received the name Stagnant Regions Model (*SRM*)[9].

The study of the Stagnant Regions Model is important in the development of geometrical characteristics of more recent trays. The use of observations of areas of stagnant liquid on trays, produced in laboratory tests, has given rise to a new generation of trays, seeking to minimize and even eliminate the regions of stagnation.

The model postulates that the flow of liquid on the tray occurs by the mechanism of the flow being plugged in the center of the tray with zones of stagnation next to the walls of the columns. The stagnant zones are defined in a simplified fashion by the areas between the column wall and the chords that connect the intake end of the downcomer and the exit of the downcomer.

In crossflow trays, when the liquid emerges from the side downcomer in a divergent section severe channeling occur. In the backmixing model, the effect of channeling is not taken into account, since the velocity profile of the liquid is considered uniform and the liquid flows by the tray as though it were in a rectangular tray, not considering the effects caused by the column wall.

Channeling of liquid is usually associated with flows in diverging channels and only when the liquid flows from side downcomers to the central downcomer. In this case, the liquid enters a divergent region and has little tendency to move to the sides in order to follow the curve of the column wall. The liquid tends to flow through the preferred channel, which is the shortest distance between the two downcomers, with regions of low velocity and stagnation on the sides of the trays. As described by Lockett[15], as there is no flow of liquid in regions of stagnation, the liquid quickly reaches equilibrium with vapor, causing the vapor that passes through these regions not to change its composition any further. Since the aim of the tray is to change the composition of the vapor, there is a reduction of tray efficiency.

Lockett[15] observed that the aforementioned considerations are simplifications and that Porter et al.[19] demonstrated that, for trays with small diameters, there is a replacement of the light components removed by the vapor crossing the stagnant regions. However, this effect occurs over a limited distance, in which the maximum width of the regions of stagnation is less than 0.5 m, regardless of the diameter of the tray. Therefore, the larger the diameter of the tray, the more significant the effect of stagnation of liquid in tray efficiency tends to be.

In the case of one-pass trays, it is expected that zones of stagnant liquid cause a significant reduction in performance, especially in large-diameter columns containing a large number of trays. This is caused because the stagnation regions are aligned one on top of another, and therefore the vapor can go through several trays without changing its composition.

Columns with two-pass trays have the advantage of reducing the effects of stagnant liquid zones. Composition of the vapor is changed by alternating the geometry of the trays, which produces an increase in efficiency, offsetting part of the efficiency lost by the tray having a shorter path length of the liquid.

Lim et al.[14] showed that maximum efficiency for one-pass crossflow trays is achieved with a column diameter of approximately 2.1 meters and for two-pass trays it should be around 2.7 m.

Porter et al.[19] presented a flow model for a one-pass crossflow valve tray in which it was assumed that the liquid that emerges from the downcomer flows through the tray directly to the downcomer outlet without spreading to the sides.

The flow pattern, similar to the one adopted by Lim et al.[14], is shown schematically in Figure 2.1. In Region I, the liquid flows through the tray at a uniform speed. But the liquid of Region II is stagnant and is not renewed by the liquid that crosses the tray.

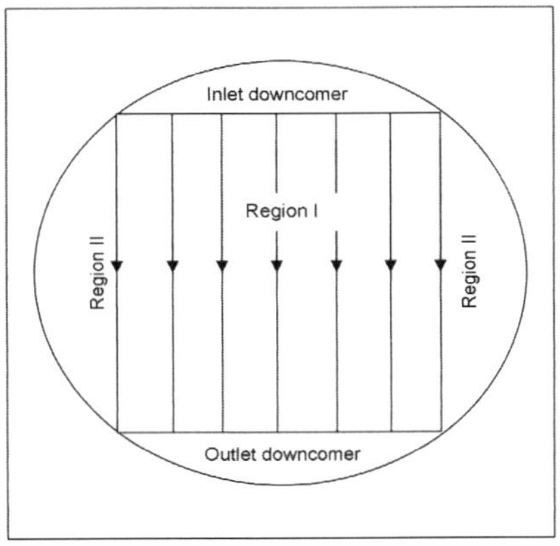

Figure 2.1: The liquid flow pattern considered by *Porter et al.*[19]

Superposed over the flow standard, there is a random movement of the particles of liquid caused by the bubbling of the vapor, which causes back mixing in Region I and an exchange of liquid between Regions I and II.

According to Porter et al.[19], for a large-diameter tray, the Murphree efficiency can be calculated by Equation 2.5.

$$\left[\frac{E_{MV}}{E_{OG}}\right]_L = \frac{A_I}{A_A}\cdot\left[\frac{E_{MVI}}{E_{OG}}\right] \tag{2.5}$$

where A_A is the active (bubbling) area on the tray and A_I is the area of region I of the tray.

For calculating the modified Murphree tray efficiency (E_{MVI}), Porter et al.[19] suggested using models of liquid mixing on the tray by Lewis[13] with corrections for calculating the number of Peclet and of λ.

In calculating the area of the region of stagnant liquid, the area of a trapezium was adopted, having as the smallest side the width of the inlet downcomer and as the largest side the result of the correlation that allows the calculation of effective width of the outlet downcomer of the trays, which was proposed by Lockett and Banik[15].

$$\frac{L_{WE}}{D_T} = 0,5 \cdot \frac{L_W}{D_T} + \frac{\left(\frac{0,5 \cdot \pi}{180}\right) arcseno \left\{\left[1 - \left(\frac{L_W}{D_T}\right)^2\right]^{0,5}\right\}}{\left[1 - \left(\frac{L_W}{D_T}\right)^2\right]^{0,5}} \tag{2.6}$$

where L_{WE} is the tray equivalent weir length - *SRM*, L_W is the weir length and D_T is the column diameter.

In the models for calculating efficiency that have been used in this work, the impact of weeping in valve trays on tray efficiency was not included, because it was found to be very low or practically null.

Some changes have been introduced in order to better adjust the models for two-pass valve trays. The Sounders Brown constant (C_{SB}) was calculated using the correlation proposed by Kister and Haas[20], obtaining a better adjustment for the column flooding factor.

Bolles[2] has considered that valve trays, with certain exceptions, behave the same way that sieve trays do. The model that describes the physical characteristics of a valve tray is the same as the one that describes the characteristics sieve trays, except for the area of the slot (A_S), which is variable in valve trays and depends on the vapor flow rate that establishes the position of the valves.

Slot area (A_S) is the characteristic parameter of operation of a valve tray, and it is defined as the area of the vertical curtain through which vapor passes out of the valve in a horizontal direction.

The model proposed by Bolles[2] was adopted in order to estimate the fraction of open valves (V_O) of valve trays, in accordance with Equation 2.7 below.

$$V_o = \frac{u_{HD} - u_{CBP}}{u_{CBO} - u_{CBP}} \qquad 0 < V_O < 1 \tag{2.7}$$

where u_{HD} is the velocity of vapor through the holes, u_{CBP} is the velocity of vapor through the holes with balance point closed, and u_{CBO} is the vapor velocity through the holes with balance point open.

For calculating the relationships between the dimensions of the tray used to convert a two-pass tray to a one-pass tray[11], the equations used were those suggested by Caldas[3] and given by the equations 2.8 and 2.9.

$$A_D = (D_T/2)^2 \arccos\left(\frac{D_T/2 - H}{D_T/2}\right) - (D_T/2 - H) * \sqrt{D_T \cdot H - H^2} \tag{2.8}$$

$$L_W = 2 \cdot \sqrt{(D_T/2)^2 - (D_T/2 - H)^2} \tag{2.9}$$

where A_D is the downcomer area, H is the downcomer length, D_T is the column diameter and L_W is the weir length.

3 Thermodynamics

The choice of a thermodynamic model is the first and most important step for obtaining accurate results in a simulation of a distillation column. With the aim of identifying which thermodynamic package best represents the test data of the column, the ones that were selected and tested were the state equation model *SRK, Peng-Robinson, RK-Soave* and the model of activity coefficient *NRTL* with the state equation *RK* (*NTRL-RK*), using the database *Ethylene* and *Pure 12*, available in the simulator *Aspen Plus 2006*. The method of activity coefficient of a liquid *NRTL* (*Non–Random Two-Liquid*) associated with the state equation *Redlich-Kwong* (*RK*) for the vapor phase – *NRTL-RK,* with the *Pure 12* database was the one that best represented the compositions and temperatures data collected in the tests performed in the plant, with emphasis on adjusting the bottom composition of the butadiene fractionator.

The NRTL activity coefficient model describes the liquid-liquid equilibrium solutions of strongly non-ideal solutions. The binary parameters were obtained from the *Dortmund*[21] database, which were already included in the database of the *Aspen Plus 2006* simulator. The *NRTL* method can deal with combinations of polar and non-polar compounds with high non-ideality.

Figure 3.1 shows the comparison of mole fractions of 1,3 butadiene in the binary mixture of 1,3 butadiene and cis-2-butene, in liquid phase, considered as a light and heavy key[20] , calculated by Aspen Plus 2006 simulator and the experimental data available in the Dortmund database[22]

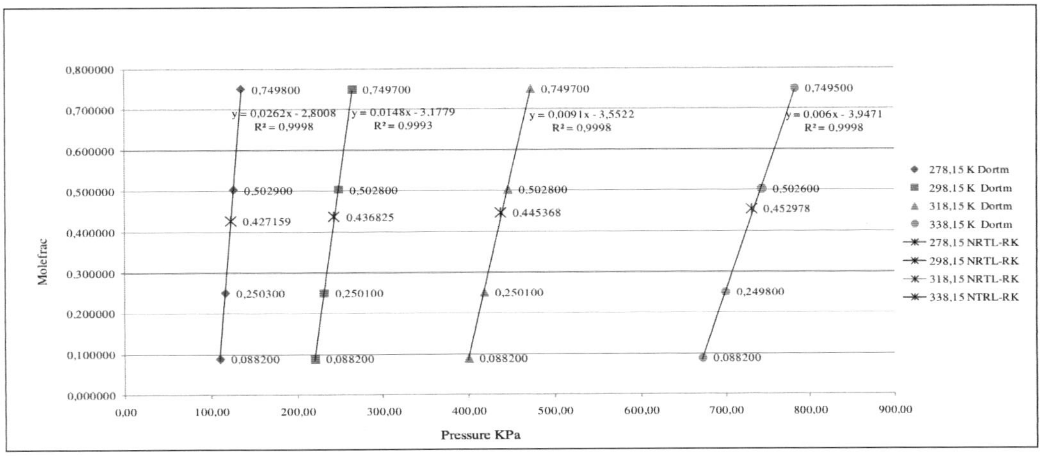

Figure 3.1: Mole fractions in liquid phase of 1,3 butadiene in the binary mixture of 1,3 butadiene and cis-2-butene calculated by *Aspen Plus 2006* simulator and the experimental values available in the *Dortmund database*[22].

The deviations of experimental data and calculated values by *Aspen Plus* simulator in the operating range of the fractionating column, 500 to 620 kPa, are less than 3.5% and are shown in the Table 3.1 below.

Table 3.1: Deviations of the experimental data – mole fraction of 1,3 butadiene in the binary mixture

Temperature	NRTL-RK Pure 12 Aspen 2006	Dortmund DB experimental	Deviation
K	molefrac	molefrac	%
278,15	0,427159	0,420259	1,615
298,15	0,436825	0,423519	3,046
318,15	0,445368	0,431411	3,134
338,15	0,452978	0,437148	3,495

The models and correlations were systematized in the *Aspen Plus 2006* simulator. An interactive algorithm was developed, combining all the models and correlations chosen, along with the geometry of the trays and the physical properties and operating conditions of the column. The algorithm was implemented in the *Aspen Plus 2006 simulator*, using Fortran 77 programming language, in the *calculator block*, allowing the calculation of apparent Murprhee tray efficiency, Murphree tray efficiency, overall efficiency of the column, and the compositions of the top and bottom for several models and correlations, and it is summarized in Figure 3.2 above.

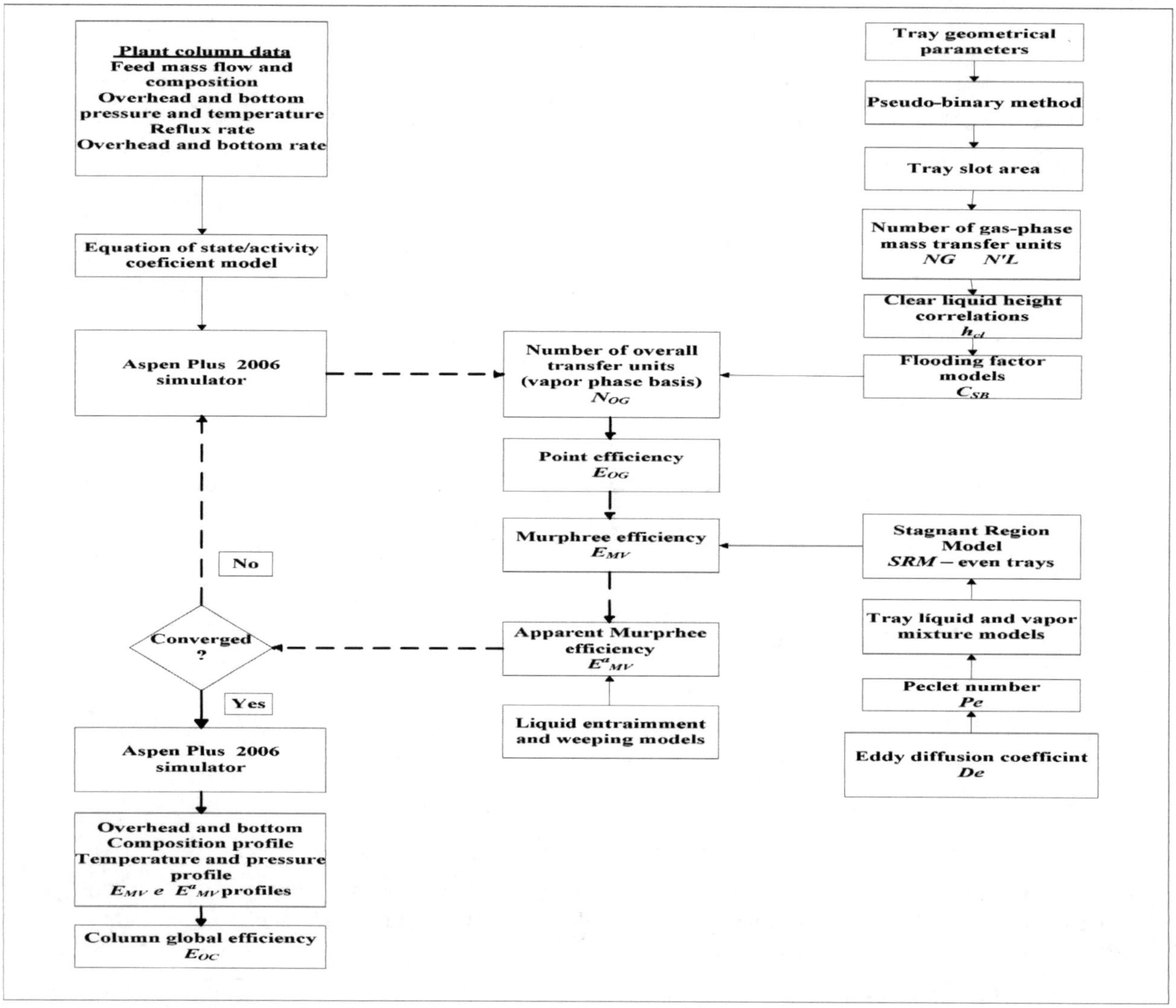

Figure 3.2: Fortran 77 algorithm implemented in Aspen Plus 2006 to calculate Murphree efficiency (E_{MV}), apparent Murphree efficiency (E^a_{MV}), and overall column efficiency (E_{OC}), based on vapor phase.

4 Column plant test

For validating the efficiency prediction models, the data collected in a test plant of a fractionating column for butadiene was used. The column was modeled in the *Aspen Plus 2006* simulator, aimed at obtaining the Murphree efficiency (E_{MV}) after which it was converted into the actual overall efficiency (E_{OC}), and it was compared to the efficiencies calculated by the prediction model.

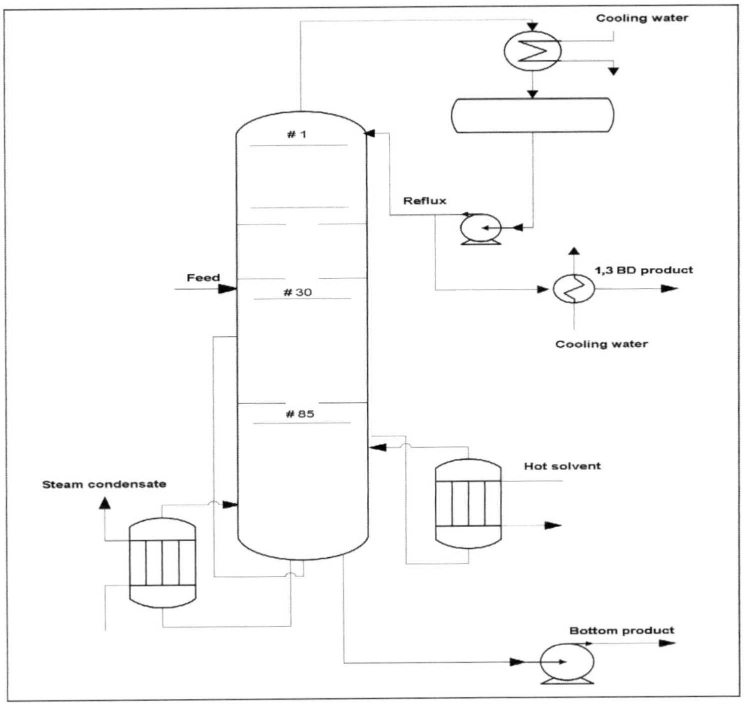

Figure 4.1: Simplified diagram of Butadiene fractionating column

The design features of the column are shown in Table 4.1:

Table 4.1: Butadiene fractionating column specifications

Number of trays	85
Feed tray	30
Trays spacings (m)	0.45
Tray type	two passes valve tray
Valve diamiter (m)	0.039-0.040
Side downcomer lenght (m)	1.35
Center downcomer lenght (m)	2.23

Comparative curves of efficiencies between trays and packing as a function of flow parameter (FP) were published by Caldas[3]. It shows a stable efficiency performance of valve trays in a wide range of operation. These results prove to be compatible with the field performance of the valve trays.

To conduct the field test, the main instruments for measuring flow, temperature and pressure were calibrated. The error in mass balance for key components was less than 5%, as recommended by Kister et al[23]. Nevertheless, to reduce the error, the production rate of butadiene was adjusted with the use of the sphere level measurement in the end storage of butadiene. Finally, the estimated error in the mass balance was + 3.5%.The operation consolidated conditions obtained through repeated field tests are shown in Tables 4.2.

.Table 4.2: Column test data

		Feed		Destillate (product)		Reflux	Bottom	
Rate	Kmol/hr	239,96		230,1		1459,27	9,898	
Temperature	°C	47,9		46,3		43,3	64,1	
Pressure	Kpa	709,02		500,14			610,95	
Heat duty	GJ/hr						32,785	
Boilup vapor	Kmol/hr						1638,46	
Composition								
1,2-Butadiene		0,69	% mole	<10	ppm mole		16,103	% mole
1,3-Butadiene		95,2	% mole	99,62	% mole		0,903	% mole
1-butane		-		-			-	
cis-2-Butene		3,4	% mole	0,303	% mole		74,927	% mole
trans-2-butene		0,135	% mole	0,079	% mole		1,42	% mole
C5+		0,103	% mole	<10	ppm mole		3,067	% mole
methylacetylene		-		6,633	% mole		-	
ethylacetylene		0,1	% mole	19,7	ppm mole		8,503	% mole
isobutene		-		0,1	ppm mole		-	
vinylacetylene		-		0,3	ppm mole		<10	ppm mole
propadiene		-		0,5	ppm mole		-	
4-TBC							-	
Amines							-	
Column temperatura profile								
Tray 1	°C	44,84						
Tray 43	°C	51,14						
Tray 57	°C	53,66						
Bottom	°C	63,16						

5 Murphree tray efficiency and Overall column efficiency

5.1 Murphree tray efficiency

The calculation of Murphree tray efficiency for two-pass crossflow trays was carried out considering the peculiarities of the odd- and even-numbered passes, and they are presented as the average tray efficiency of the column, as well as the average deviation.

For the conversion of Murphree tray efficiency of the column, resulting in the simulation to overall efficiency, Case 1 of Lewis[13] model was used.

Applying the model of Chan and Fair[4] resulted in Murphree tray efficiency (E_{MV}) of 104.6% and showed a deviation of 21.1% when compared to the efficiency of the column that was based on the plant test and adjusted in the simulator. The results of tray efficiency using the model by Garcia[6] presented a tray efficiency of 137.3%, with a deviation of 60% compared to the adjusted tray efficiencies.

Figure 5.1 illustrates the Murphree tray efficiencies obtained for each tray by using the model of Chan and Fair[4] and Garcia[6], and their fittings to point out that the differences between the efficiencies of even- and odd-numbered trays are due to differences in tray geometry, causing a "*sawtooth effect*".

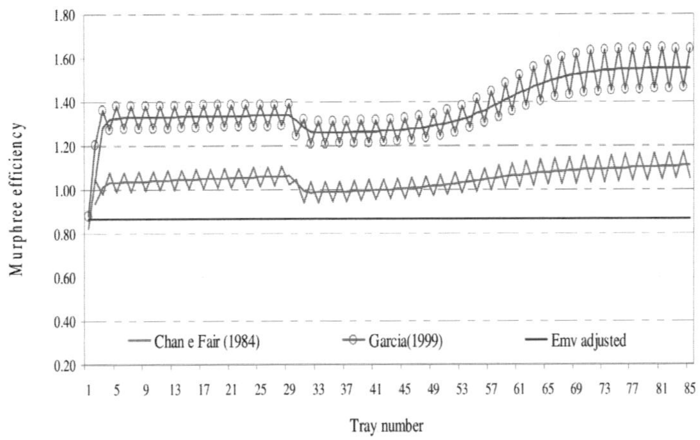

Figure 5.1: The Murphree efficiencies of Chan and Fair[4] and Garcia[6] of each tray.

The conversion from a two-pass tray to a one-pass tray proposed by Klemola and Ilme[11] causes all the plates in the column to have the same geometric configuration, resolving differences in efficiency between odd- and even-numbered trays.

By adopting the criterion of Klemola and Ilme[11], the modified model of Garcia[6] presented a tray efficiency of 134.2 % with deviations in relation to the tray efficiencies adjusted in the simulator of 55.3%.

The modified model of Chan and Fair[4] presented a tray efficiency of 91.4% with a deviation of 5.8% in relation to the efficiency adjusted in the simulator.

Figure 5.2 illustrates the tray efficiencies obtained from each tray by the models of Chan and Fair[4] and Garcia[6], using the equivalent one-pass tray liquid flow path length (C_E) of Klemola and Ilme[11].

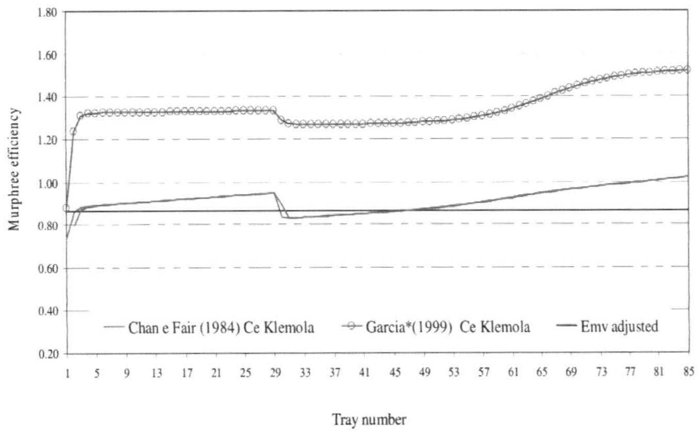

Figure 5.2: Efficiencies of Murphree, Chan and Fair[4] and Garcia[6] C_E of Klemola and Ilme[11] of each tray.

5.2 Murphree tray efficiency – models selected by using the *SRM* model

In this section, it was present the results of the Murphree tray efficiencies (E_{MV}) and the overall efficiency of the column (E_{OC}) by using the proposed models adjusted in accordance with the Stagnant Region Model (*SRM*).

As the correlation by O'Connell[17] does not take into account geometric parameters of the tray, but only the relative volatility and viscosity of the feed, taken as the arithmetic average temperature of top and bottom of the column, it is not fitting to use the *SRM* model for this correlation.

In cross-flow two-pass trays, the liquid emerges from the downcomer in a divergent section it flows from the side into to the center of the downcomer, causing severe channeling. This fact causes a reduction in efficiency by unsuitable flow of liquid on the plate.

In the column evaluated, only in even-numbered trays does liquid emerge from the downcomer in a divergent section, and therefore the *SRM* model is only applied to these trays.

The calculation of the efficiency of even-numbered trays considers the length of the central downcomer modified by the *SRM* model, while for the calculation of odd-numbered trays the design length of the side downcomer was considered.

In order to compare the results of the Murphree tray efficiencies (E_{MV}) obtained by using the models of Chan and Fair[4], Garcia[6], and the modified models of Chan and Fair[4] and Garcia[6], by using the concept of equivalent length (C_E) the efficiencies were calculated using the *SRM* model.

The efficiencies calculated with the model of Garcia[6] and with the modified model of Garcia[6], which considers equivalent length (C_E) by using the *SRM* model were 113.75% and 108.52%, respectively. The deviations from Murphree tray efficiency adjusted by the simulator were reduced when compared to the same efficiencies without the use of the *SRM* model, but the results still overestimate the value of efficiencies.

The results with the model of Chan and Fair[4] had a Murphree tray efficiency of 96.2% and a deviation from the adjusted efficiency of 11.3%, showing that the deviation of the Murphree tray efficiency calculated by using the *SRM* method was also reduced. In the same way as the model of Garcia[6], the result tends to overestimate the value of efficiencies, albeit considerably smaller.

Figure 5.3 illustrates the Murphree efficiencies obtained for each tray by using the model of Chan and Fair[4] and Garcia[6] by using the *SRM* model.

The efficiencies calculated by the model of Garcia[6] using the SRM present a larger difference between efficiencies of even-numbered trays when compared to the results of the same model without using the *SRM.* These differences are explained by the formation of stagnant regions in the even-numbered trays, which has the flow of liquid from the side toward the center, thus reducing the efficiency of these trays.

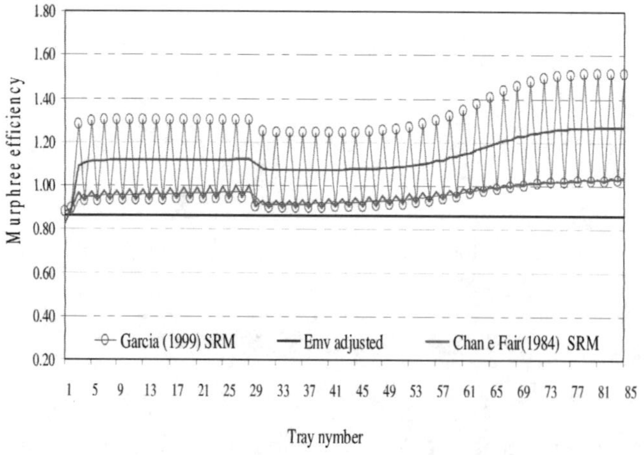

Figure 5.3: Chan and Fair[4] and Garcia[6] Murphree tray efficiencies with the use of the *SRM* model.

Figure 5.4 illustrates the Murphree efficiencies obtained for each tray by using the models of Chan and Fair[4] and Garcia[6] by using the C_E proposed by Klemola and Ilme[11], together with the *SRM* model. The "*sawtooth effect*" is not present now that the trays of the column have the same geometric configuration.

The use of the modified model of Chan and Fair[4], considering the equivalent length C_E and using the *SRM* model resulted in a Murphree tray efficiency of 75.3% with a deviation from the adjusted efficiency -12.9% underestimating the efficiency.

The results of the Murphree tray efficiency applying the model modified by Garcia[6], which considers the equivalent length C_E using the *SRM* model was 108.5% with a deviation from the adjusted efficiency of 25.6%.

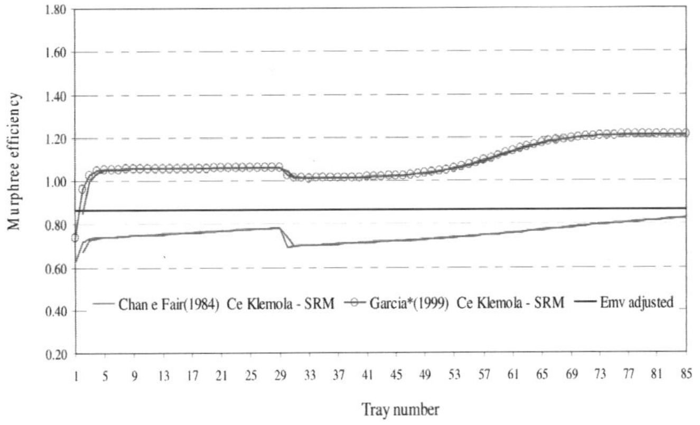

Figure 3.4: Murphree tray efficiencies from Chan and Fair[4] and Garcia[6] using C_E from Klemola and Ilme[11] and the *SRM* model.

5.3 Overall column efficiencies

Figure 5.5 shows the comparison of overall column efficiencies (E_{OC}) calculated by several models in relation to the overall efficiency adjusted in the simulator for the butadiene fractionating column. Lines of variation of ±10%, were allowed, which shows an acceptable deviation related to the number of trays in the column.

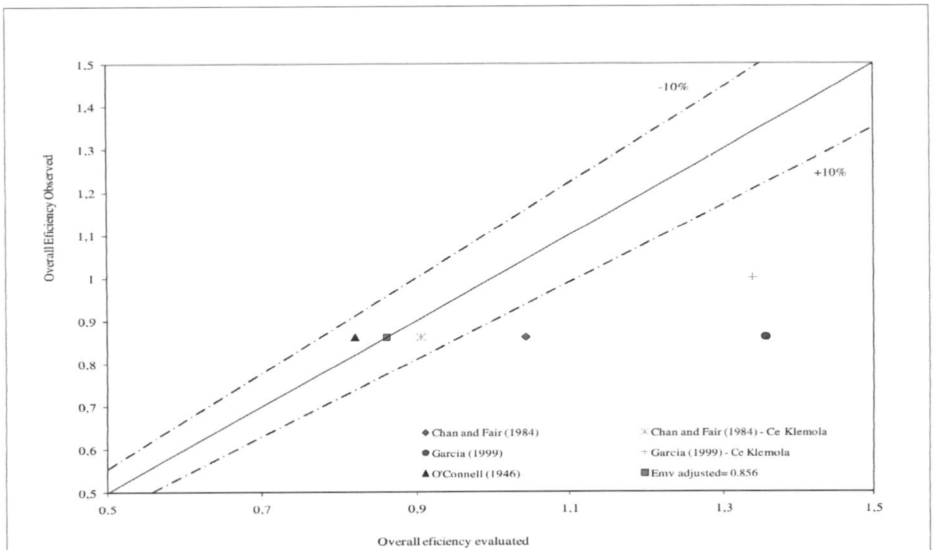

Figure 5.5: Comparison of predicted overall efficiencies (E_{OC}) related to the overall efficiency adjusted in the Simulator for a butadiene fractioning column – without the *SRM* model.

It is verified in Figure 3.5 that the overall efficiency (E_{OC}) calculated by the modified model of Chan and Fair[4], using the equivalent length recommended by Klemola and Ilme[11] and by the correlation proposed by O'Connell[17] presented the lowest deviations.

The overall efficiency calculated by the model of Chan and Fair[4] was shown not to be conservative presenting a positive deviation of 21.0% compared to the adjusted efficiency.

The results of the overall efficiencies (E_{OC}) obtained by using the model of Garcia[6] and the modified model of Garcia showed a deviation related to the adjusted efficiencies of 57.4% and 55.4%, respectively, which overestimates the values of these efficiencies and makes it unadvisable to use these methods for calculating the overall efficiency of two-pass crossflow trays.

5.4 Overall column efficiencies – models selected by using the *SRM* model

Figure 5.6 shows the comparison of overall efficiencies (E_{OC}) calculated by several models in relation to the overall efficiency adjusted in the simulator for the butadiene fractioning column using the *SRM* model.

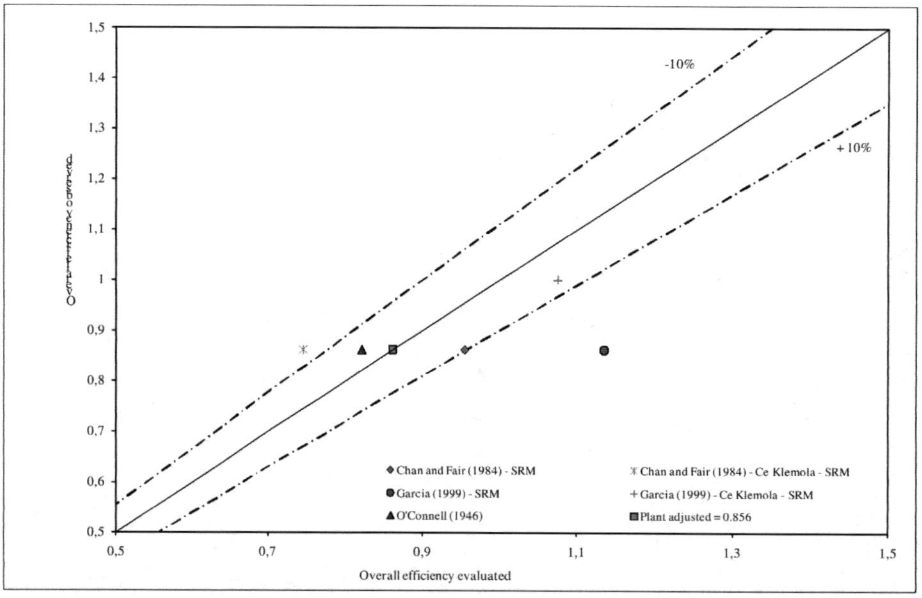

Figure 5.6: Comparison of overall efficiencies (E_{OC}) related to the overall efficiency adjusted in the Simulator for the column tested – using the *SRM* model.

By observing Figure 3.6, it is possible to verify that the values of overall efficiency of the column (E_{OC}) calculated by the model of Chan and Fair[4] by using the *SRM* model and by the correlation of O'Connell[17] present the lowest deviations.

The overall efficiency calculated by the model of Chan and Fair[4] with the *SRM* proved to be non-conservative, showing a positive deviation of 10.9% over the adjusted efficiency.

The results of the overall efficiencies of the column (E_{OC}), obtained by using the model of Garcia[6] and the modified model of Garcia using the *SRM* model were shown to be more adjusted than the same models without using the *SRM*, presenting a deviation from the adjusted overall efficiency of 31.7% and 24.5%, respectively, which overestimates the values of these efficiencies.

The modified models of Chan and Fair[4] and Garcia[6] by using C_E and the *SRM* model showed no acceptable adjustments to the composition of the 1.3 butadiene in the bottom of the column, which invalidates their use in calculating the overall efficiency (E_{OC}) of the column tested.

6 Conclusions

The simultaneous transfer of heat and mass, combined with the complex hydraulics of fluid in two phases makes distillation a vast field for analysis of its fundamentals. Distillation had become a well-established operation long before modern theories of mass transfer were consolidated. Therefore, the common tendency of most research in distillation still maintains its empirical, semi-empirical, or mechanistic nature. It is in this scenario that the conclusions of this work were obtained.

The correlation of O'Connell[17] showed itself to be well-adjusted to the overall efficiency of a column (E_{OC}), which supports its adoption as a standard in the industry yet today.

Since the correlation does not take into account the hydraulic factors of trays, no information is obtained in relation to the phenomena that occur in the mass transfer in the trays and adds little to the understanding of the key factors of the current designs of trays and little or nothing to the development of new models.

The model of Chan and Fair[4] showed deviations in the Murphree tray efficiencies within the limits set by the very authors of the method.

The introduction of the *SRM* (*Stagnant Regions Model*) model to the model of Chan and Fair[4] led to lower deviations, but the results still overestimate the values of the Murphree tray efficiencies.

The model of Chan and Fair[4], associated with the model proposed by Klemola and Ilme[11] regarding the conversion of two-pass trays to one-pass trays, keeping the active area of bubbling (A_A) in a two-pass tray, suitably represented the efficiency of the column tray studied, validating the use of this model as an alternative for rapid and easy deployment.

The model of Garcia[6], with or without the use of the *SRM* model, showed significant deviations of tray efficiency when compared with the adjusted efficiency of the column that was tested, which shows that it is necessary to further develop in this model. However, it is worth noting that the calculation model of point efficiency (E_{OG}) of Garcia[6] with and without the use of the *SRM* model deepens the knowledge hither to available for mass transfer phenomena that occur on a tray.

In summary, the correlation of O'Connell[17] was the one that best represented the overall efficiency of the column tested and is recommended to estimate the overall efficiency of columns equipped with two-pass valve trays. The model of Chan and Fair[4] combined with *SRM* model and the model of Chan and Fair[4] with the use of equivalent length (C_E) proposed by Klemola and Ilme[11] slightly overestimated the overall efficiency of the column. However, this difficulty can be circumvented by using a correction factor that takes into account the calculation of the number of actual trays in the column in the deviations presented by both methods.

Finally, it is emphasized that the findings found here are based on studies of a single column and its representation must be taken into account with due caution.

List of acronyms

A_S valve try slot area (m^2)
A_A active (bubbling) area on the tray, m^2
A_I area of the region I of the tray, m^2, eqn. (2.5)
C_E equivalent one-pass tray liquid flow path length, m
C_{SB} Saunders Brown constant, m/s
De liquid-phase eddy diffusivity, m^2/s
D_T column diameter, m
E_{OC} overall column efficiency
E_{OG} point efficiency
E_{MV} Murphree vapor phase tray efficiency
$E^a{}_{MV}$ apparent Murphree vapor phase tray efficiency
E_{MVI} Murphree vapor phase tray efficiency for region I, eqn (2.5)
G' vapor flow rate per unit bubbling area, kmol / (m^2.s)
h_{CL} clear liquid height, n
L_{WE} downcomer effective width one-pass tray, m, eqn. (2.6)
L' liquid flow rate per unit bubbling area kmol / (m^2.s)
L_W weir length, m
M_G vapor flow rate, kg/s
M_L liquid flow rate, kg/s
M_{WG} vapor mixture molecular weight, kg/kmol
M_{WL} liquid mixture molecular weight, kg/kmol
m slope of equilibrium line (binary), dy/dx
N_G number of binary vapor phase transfer units
N'_L number of binary liquid phase transfer units
N_{OG} number of overall binary vapor phase transfer units
FP flow parameter
u_{HD} velocity of vapor through the holes, m/s
u_{CBP} velocity of vapor through the holes, balance point closed, m/s
u_{CBO} vapor velocity through the holes, balance point open, m/s.
V_O fraction of open valves
λ ratio of slope of equilibrium line (m) to slope of operating line (L'/G'):
y vapor molar fraction
$\overline{y}_n$ value of y in vapor leaving tray n
$y_n{}^*$ mole fraction in vapor in equilibrium with liquid n
Pe Peclet number

5. References

1. American Institute of Chemical Engineers. *Bubble Tray Design Manual.* New York:The Science Press, Inc., 1958. 96 p.

2. Bolles, W. L. Estimating Valve Tray Performance. *Chem. Eng. Progress*, 1976; 43-50.

3. Caldas, J.N., Lacerda, A.I., Veloso, E., Paschoal, L.C.M. *Internos de Torres.* Rio de Janeiro: Interciência, 2007. 509 p.

4. Chan, H., Fair, J. R. Predict of Point Efficiencies on Sieve Trays. 1. Binary Systems. *Ind. Eng. Chem. Process Des. Dev.*, 1984; 23(4):814-819.

5. Domingues, T.L., Secchi, A.R., Mendes, T.F. Overall Efficiency Evaluation of Commercial Distillation Columns with Valve and Dualflow Trays. AIChE J., 2010; 56(9):2323-2330.

6. Garcia, J.A. *Fundamental Model for the Prediction of Distillation Sieve Tray Efficiency: Hydrocarbon and Aqueous Systems.* Austin: The University of Texas at Austin, 1999. PhD. Thesis, The University of Texas at Austin, 1999.

7. Garcia, J.A., Fair, J.R. A Fundamental Model for the Prediction of Distillation Sieve Tray Efficiency. 1. Database Development. *Ind. Eng. Chem. Res.*, 2000; 39(6):1809-1817.

8. Invernici, P.L. *Avaliação da eficiência de pratos valvulados de dois passes com downcomer de uma coluna de destilação industrial*, M.Sc. Thesis in Chemical Engineering (in Portuguese), Porto Alegre: Universidade Federal do Rio Grande do Sul, 2009.

9. Kister, H.Z. Distillation Design. New York: McGraw-Hill Inc., 1992. 710 p.

10. Klemola, K.T., Ilme, J.K. Calculation Methods for the Numbers of Transfer Units on Sieve Trays. In: *AIChE Annual Meeting*, 1997a, Los Angeles.

11. Klemola, K.T., Ilme, J.K. Effect of Two-Pass Trays on Distillation Efficiencies. *Chem. Eng. Technol., v. 20, n. 7, p. 478-84, 1997b.*

12. Kirchbaum. F. *GebieteIngenieurw.,* 1934; 5:245-52.

13. Lewis,W.K. Rectification of Binary Mixtures. *Ind. Eng. Chem. Res.*, 1936; 28(14):339-402.

14. Lim, C.T., Porter, K.E., Lockett M.J. The Effect of Liquid Channelling on Two-Pass Distillation Plate Efficiency *Ind. Eng. Chem. Res.*, 1974; 52:193-201.

15. Lockett, M. J. Distillation Tray Fundamentals. Cambridge: Cambridge University Press, 1986. 226 p.

16. Caldas, J.N., Lacerda, A.I., Veloso, E., Paschoal, L.C.M. Internos de Torres. Rio de Janeiro: Interciência, 2007. 509 p.

17. O'Conell, H.E. Plate Efficiency of Fractionating Columns and Absorbes, *Trans. Inst. Chem. Engrs*, 1946; 42: 741-755.

18. Prado, M., Johnson, K. L., Fair, J. R. Bubble-to-Spray Transition on Sieve Trays. *Chem. Eng. Progress*, 1987; 32-40.

19. Porter, K.E., Lockett, M.J., Lim, C.T. The Effect of Liquid Channelling on Distillation Plate Efficiency. *Trans. Inst. Chem. Engrs*, 1972; 50:91-101.

20. Kister, H.Z., Haas, J.R. *Chem. Eng. Progress*, 1990; 86(9):63.

21. Aspen Technology Inc. Physical Property Methods and Models: release 10.2. Manual. Cambridge, EUA: Aspen Technology Inc., 2000.

22. Gmehling, J. Vapor-Liquid Equilibria. Frankfurt: *Dortmund Data Base*, 2011.Reference number 1566. (Experimental data)

23. Kister, H. Z; Brown, E.; Sorensen, K. *Sensitivity analysis is key to successful DC5 simulation. Hydrocarbon Processing*, 1998;124-132.

Predicting Distillation Tray Efficiencies Based on Modified O'Connell Correlation Analysis (MOCA) Technique

Steven G. Schon, P.E.
Engineering Fellow
Technology and Engineering
Arkema Inc.
900 First Avenue
King of Prussia, PA 19406

ABSTRACT

A methodology is presented wherein the tray efficiencies can be estimated from actual column temperature profiles compared with those predicted by simulations. The technique is based on a modification to the well-established O'Connell correlation for overall column efficiencies, using correction factors applied to the basic O'Connell correlation. The methodology has been used successfully to assess the efficiencies of existing columns, and for the design of new columns with minimal over-design and associated costs.

It is proposed to develop a database and correlations for correction factors to predict the individual stage (Murphree) efficiencies of various tray designs used in distillation columns. Such a database would have broad applicability to distillation in the refining and chemical industries, leading to more success and cost-effective design and column operations,

BACKGROUND

In this era of computational and simulation tools for distillation, it is relatively easy to calculate the required number of theoretical stages, reflux, and optimal feed location to achieve a specified separation. However, it remains a significant challenge to translate theoretical stages into design for real-world trayed distillation columns, i.e. specifying the number of trays and feed point. The chief hurdle is predicting the tray efficiencies, especially as the efficiencies may vary substantially at different locations in a column, is there are wide variations in relative volatility or transport properties. Similarly, it is difficult to estimate the (localized or point) tray efficiencies of existing columns, do the frequent lack of sampling points, as well as the hazards associated with sampling liquids and vapors from an operating unit.

CURRENT APPROACH FOR PREDICTING TRAY EFFICIENCIES

While a variety of theoretical models and empirical correlations have been proposed to predict tray efficiencies based on tray geometry considerations and fluid physical properties, to date, there is no satisfactory general method for predicting tray efficiencies for specific column designs.

The majority of tray efficiency data reported in the literature are based on overall column efficiencies (E_{OC}), i.e. the ratio of number of theoretical staged required for the separation (excluding the reboiler and condenser) to the actual number of trays in the tower. E_{OC} is typically calculated by matching the number of theoretical stages in stage-to-stage calculations, e.g. from tower simulations, to actual column results at known operating conditions (pressures, compositions, reflux ratios, etc.).

As Kister points out in his book Distillation Design[1], "several empirical tray efficiency correlations are available in the literature. Of these, the O'Connell correlation has been the standard of the industry for several decades". The O'Connell correlation, published in 1946 and pre-dating the era of advanced tray designs, was based on test data from 31 industrial columns, most of which used bubble-cap trays. It fit a curve of overall column efficiency vs. the product of relative volatility of the key components times the viscosity of the feed at average column conditions. The original O'Connell correlation for trayed distillation columns[2] is reproduced in **Figure 1.**

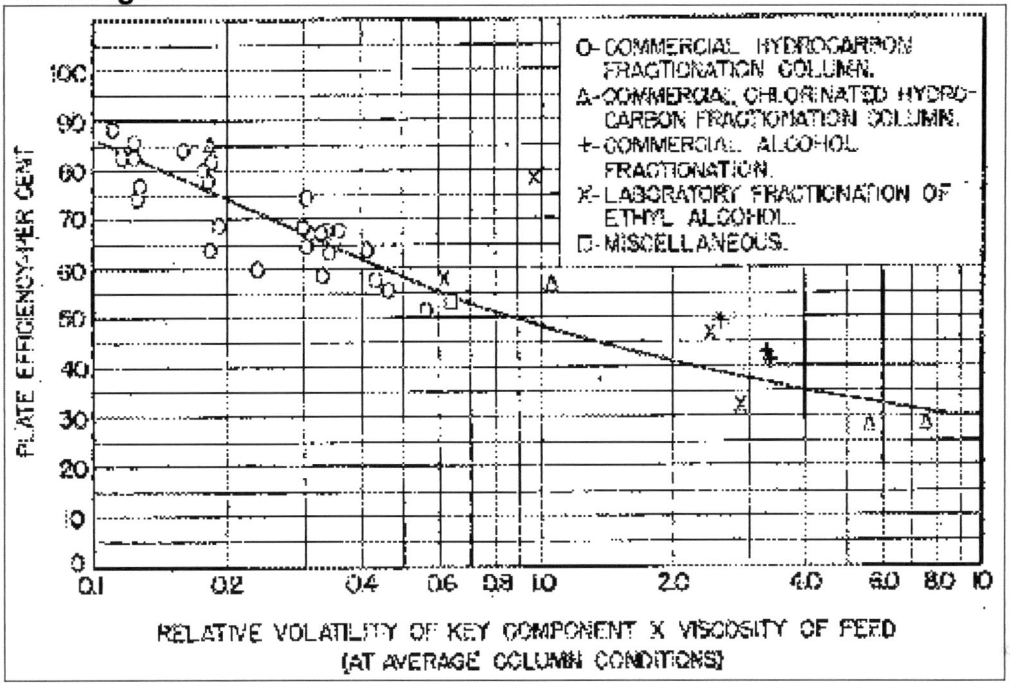

Figure 1

Note that the O'Connell curve is a semi-log plot. The curve can be linearized by converting to a log-log plot, as shown in **Figure 2**. The linearized form is:

$$\log \boldsymbol{\eta} = \mathbf{m} * \log[\mathbf{Param}] + \mathbf{b} \qquad \text{(Equation 1)}$$

where

$\boldsymbol{\eta}$ = Overall Efficiency
m = slope of log-log O'Connell correlation = **-0.244**
b = intercept of log-log O'Connell correlation = **1.6917**
Param = O'Connell Parameter
= Relative Volatility ($\boldsymbol{\alpha}$) * Feed Viscosity ($\boldsymbol{\mu}$) in cP

and

$$\boldsymbol{\alpha} = K_{light\ key} / K_{heavy\ key} = (y_{lk} / x_{lk}) / (y_{hk} / x_{hk})$$

---

1 Henry Z. Kister, "**Distillation Design**", McGraw Hill (1992), p. 376.
2 H.E. O'Connell, Trans. AIChE, **42**, p. 741

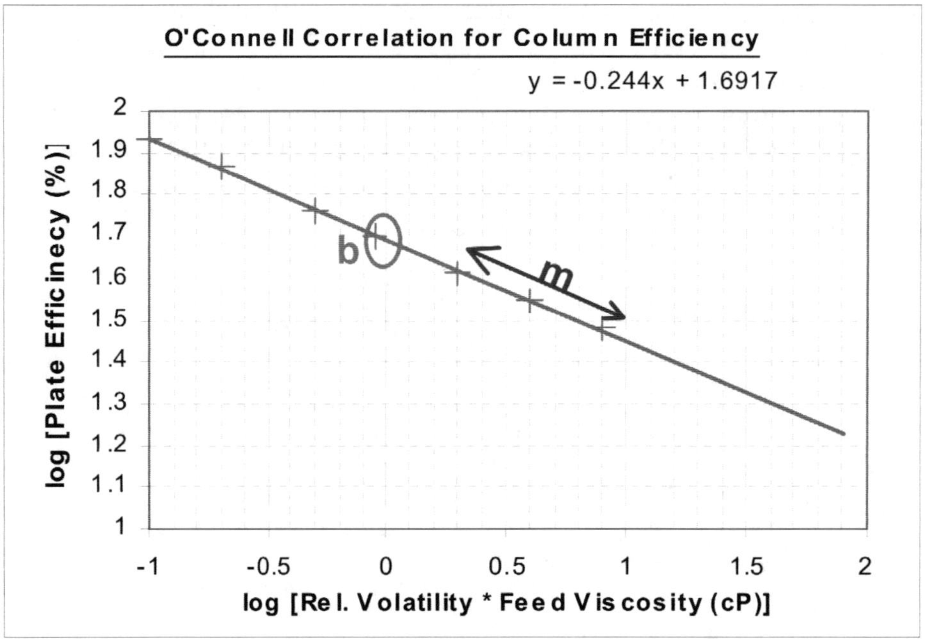

Figure 2

Taking the anti-log of equation (1), the efficiency can predicted from the O'Connell parameter and the slope and intercept of the log-log plot of Figure 2:

$$\eta = 10^{(\mathbf{-0.244} \, * \, \log[\alpha \, * \, \mu] \, + \, \mathbf{1.6917})} \qquad \text{(Equation 2)}$$

<u>MODIFIED O'CONNELL CORRELATION ANALYSIS (MOCA) TECHNIQUE</u>.

As Kister noted in a recent article[3], "since tray efficiencies may vary from one tower section to another, the overall column efficiency concept is often applied separately to the stripping section and the rectification section…This concept is, therefore, preferred by most industrial practitioners."

Modern distillation simulation software tools readily generate full stage-to-stage profiles of K-values (ratio of vapor to liquid mole fractions: $K_i = y_i / x_i$) for all components, as well as transport properties, including viscosities. It is straight-forward to calculate the O'Connell parameter (i.e. the product of relative volatility of the key components times the viscosity) for every stage in a column.

It is therefore proposed to expand the efficiency calculations, from separate column sections, to predict the efficiencies of (all) individual trays in a column, using O'Connell's approach, i.e. based on the product of key component relative volatilities and liquid viscosities. The concept is to apply the O'Connell correlation, but with correction factors that are empirically determined for specific tray geometries. This will be referred to as the Modified O'Connell Correlation Analysis (MOCA) technique.

3 Henry Z. Kister, "**Effects of Design on Tray Efficiency in Commercial Towers**", <u>Chemical Engineering Progress</u>, June 2008, pp. 39 –47.

The essence of the MOCA technique lies in applying the O'Connell correlation to predict the stage-wise Murphree efficiencies, with 2 empirically determined correction factors to the slope and intercept used in the exponential term of equation 2:

$$\eta_M = 10^{[(-0.244+a) * \log[\alpha * \mu] + (1.6917+b)]} \qquad \text{(Equation 3)}$$

$$= 10^{[(A) * \log[\alpha * \mu] + (B)]} \qquad \text{(Equation 4)}$$

where

A = slope of modified O'Connell curve = -0.244 + **a**
B = intercept of modified O'Connell curve = 1.6917 + **b**

The empirical correction factors are determined by getting the best match of the column temperature profiles predicted by a simulator (using the calculated efficiencies) against the actual column temperature profiles for a known tray geometry and operating conditions. The calculation sequence is as follows:

1. Make Initial guess of η_M for each stage. (As a starting point, use typical values for the tray types being used).
2. Run a simulation of the column, using the actual number of trays and the assumed Murphree efficiencies from step 1. From the simulation results column profiles, obtain stage-wise K values of key components & viscosities. Calculate the O'Connell parameter for each tray.
3. Assign **a** and **b** correction factor values. (As a starting point, use either zero, or values obtained by MOCA for similar tray types).
4. Re-calculate η_M for each stage, using Equation 3.
5. Re-run the simulation, using the re-calculated η_M from Step 4.
6. Get new values for K's & viscosities and re-calculate O'Connell parameters. Re-calculate η_M for each stage, using Equation 3.
7. Compare the new η_M values against those calculated in step 4. Repeat steps 4 – 6 (replace the updated η_M values from Step 6 in step 4) until the calculated stage efficiencies stabilize / converge.
8. Compare predicted vs. actual T profiles. If not a close match, try different values for A,B (step 3) and repeat steps 4 –7.

MOCA METHODOLGY - EXAMPLE

A worked-out example is illustrated in **Figures 3 and 4**.

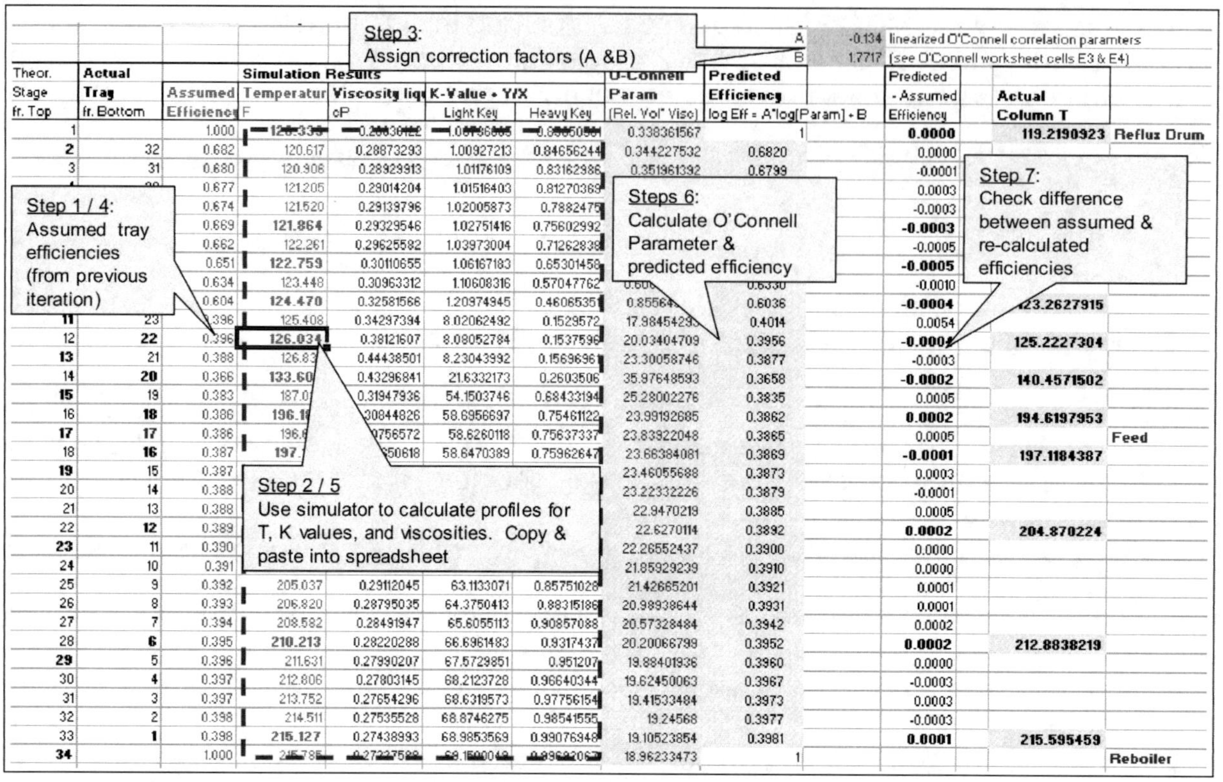

Figure 3

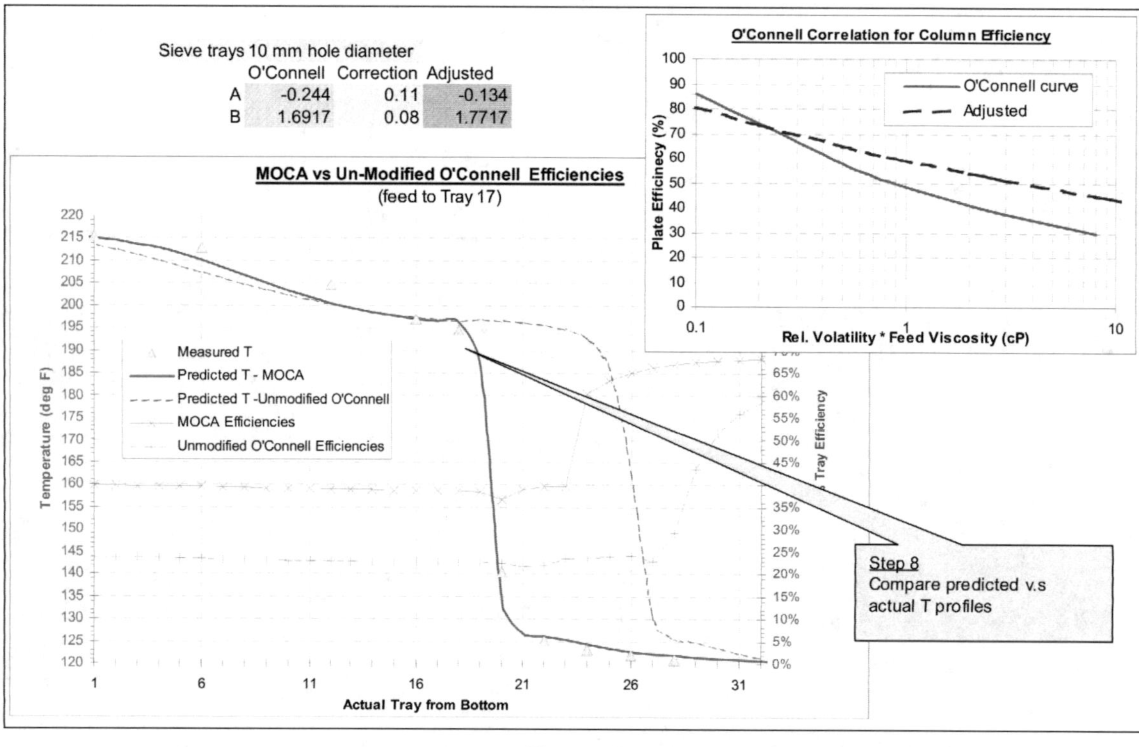

Figure 4

CONSIDERATIONS WHEN USING MOCA

For MOCA to be effective, the distillation simulations need good accuracy. This requires a representative thermodynamic model for the vapor liquid equilibria (VLE) of at least the key components, as well as reliable calculation of transport properties, especially the mixture viscosities.

It is also essential to have meaningful column temperature profiles. There should be multiple temperature measurement points, especially in zones where the composition and temperature profiles change rapidly from tray to tray.

When comparing real column data, it should be done under conditions within the normal hydraulic loading range, i.e. the zone of the so-called uniform-efficiency plateaus, typically at 50 – 80% of flood. Outside this range, weeping or incipient flooding can occur, resulting in substantially lower efficiencies.

Always bear in mind that MOCA is an empirical methodology, and that the correction factors (a, b) or equivalently, the slope and intercept (A,B) of the linearized modified O'Connell plot, are tray-geometry-specific. Even within a given tray type, a number of factors can impact tray efficiency and hence the correction factors, including path length, hole diameter, and to a lesser extent, hole area and weir height[4]. Use MOCA for scale-up or design of new columns only if correction factors are available (e.g. from existing column data) for the specific tray geometries that are planned for the new unit.

EXAMPLES

1. Calculation of existing column tray efficiencies

   A high-pressure splitter tower was modeled, and MOCA was used to back-fit the tray efficiencies. While the efficiencies differed and were substantially constant in the stripping section and most of the rectification section, it was found that there was a transition in efficiency over 8 trays, starting at about 12 trays above the feed tray (see **Figure 5**)

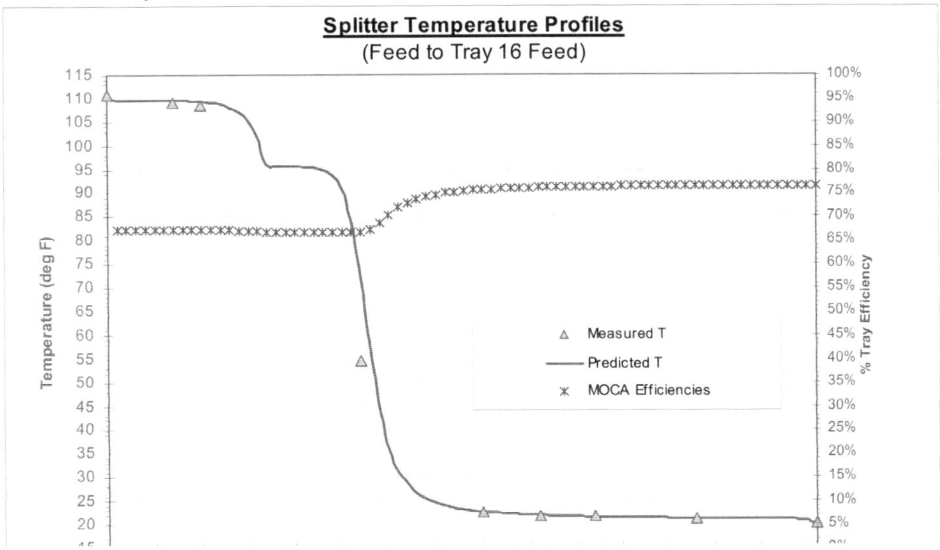

Figure 5

4 Henry Z. Kister, "**Effects of Design on Tray Efficiency in Commercial Towers**", Chemical Engineering Progress, June 2008, pp. 39 –47.

Subsequent simulations using the actual number of trays and MOCA-calculated efficiencies indicated that the column was significantly over-refluxed presenting an opportunity to reduce energy consumption. However, after the plant reduced the boil-up, the separation was significantly worse than predicted, and there was a marked shift in the temperature profile below the feed. The MOCA analysis was repeated, however, the only way to get the temperature profiles to match the plant data was to assume very low efficiencies (< 10%) below the feed. This led us to suspect the stripping section was weeping; this was confirmed when the hydraulics were checked – at the reduced rates, the stripping section was calculated to be operating below 30% (see **Figure 6**). The stripping temperature profile discrepancies were apparent, albeit reduced, at intermediate reboiler duties, based on hot oil (H.O.) flows (see **Figure 7**).

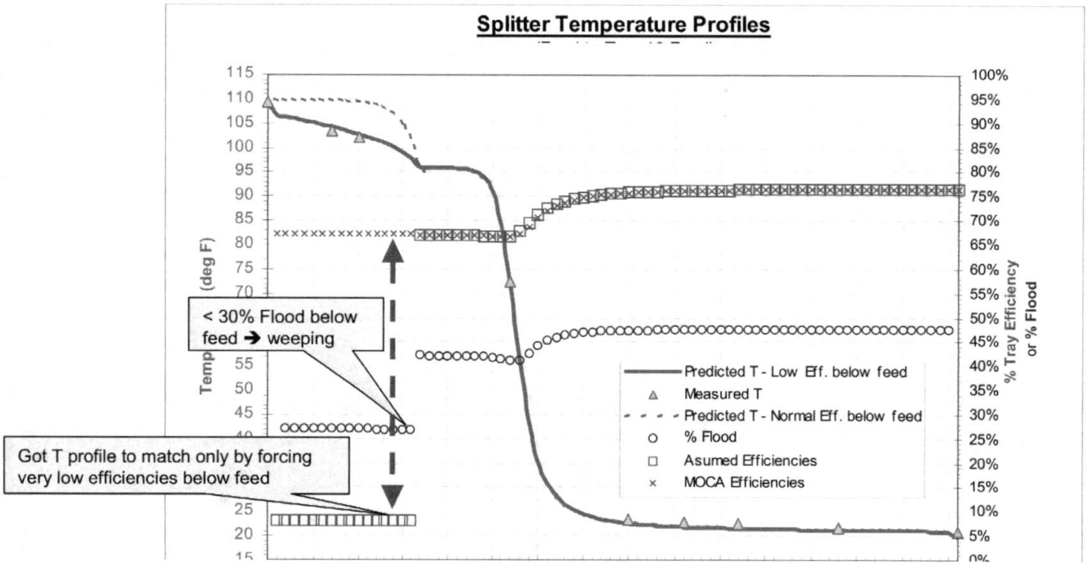

Figure 6

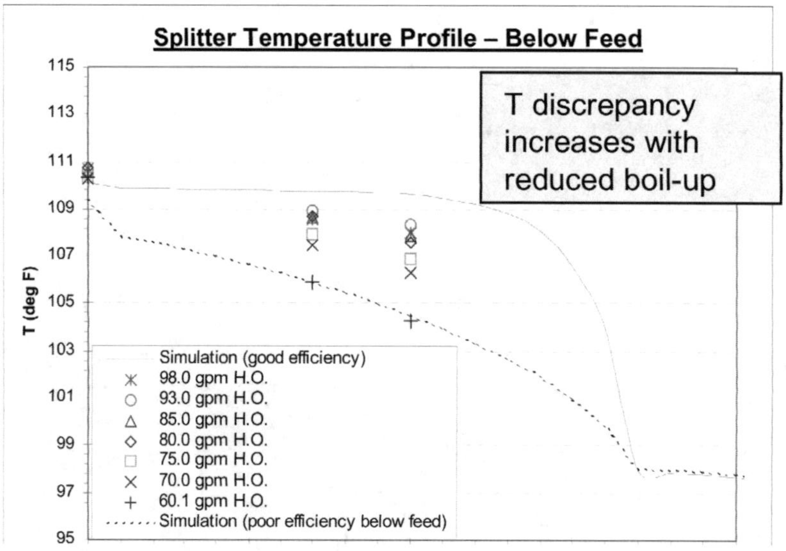

Figure 7

2. <u>Verifying Correction Factors Insensitive to Components Being Separated</u>

A small pilot plant column was used with a reference binary system (acetone-water), and the MOCA technique was used to determine the correction factors for the pilot column trays. The predicted and measured temperature profiles and efficiencies are shown in **Figure 8**.

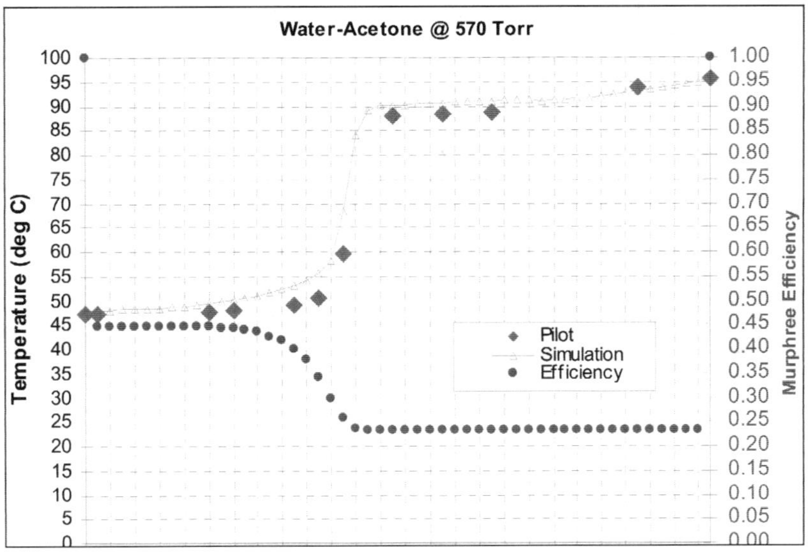

Figure 8

Another pilot column was built, using the same tray geometry, but separating a proprietary ternary polar mixture (A-B-water), and set-up as a low-pressure side-draw system. Using the same correction factors from the acetone-water data the simulation-predicted temperature profiles closely match the experimental pilot profiles, as shown in **Figure 9**. It should be noted that for this system, different pairs of light and heavy keys were used above and below the side-draw, accounting for the step-change in efficiencies.

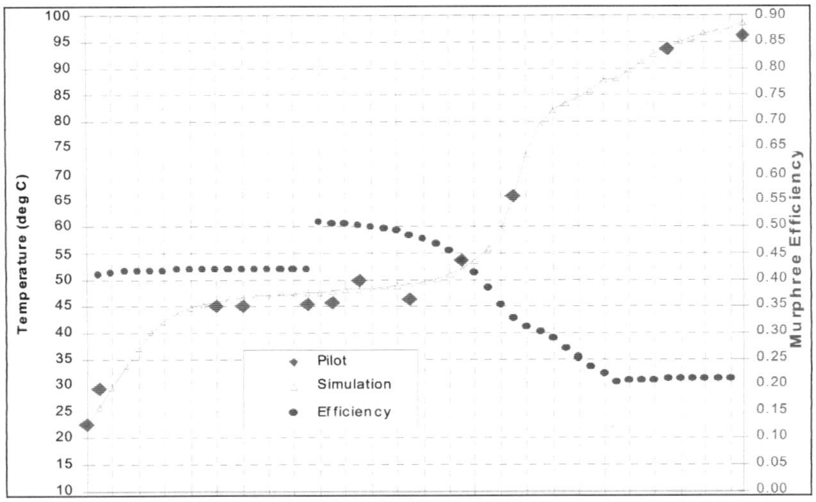

Figure 9

These results gave us confidence that the MOCA-predicted efficiencies were substantially independent of the specific components being distilled, and that the O'Connell slope/intercept correction factors were primarily dependent on the tray geometries.

3. Design of New Column Using Efficiency Factors from an Existing Column

The (a,b) correction factors and the resulting modified O'Connell plots for the pilot and industrial columns described in Example 2, are compared in **Figure 10**. The pilot column averaged about 20% lower (absolute) efficiency compared to the industrial unit. It is interesting to note that both trays had the same "b" values; it is not known at this time if this is merely coincidence, or whether it arises because of some common feature (e.g. sieve tray configuration).

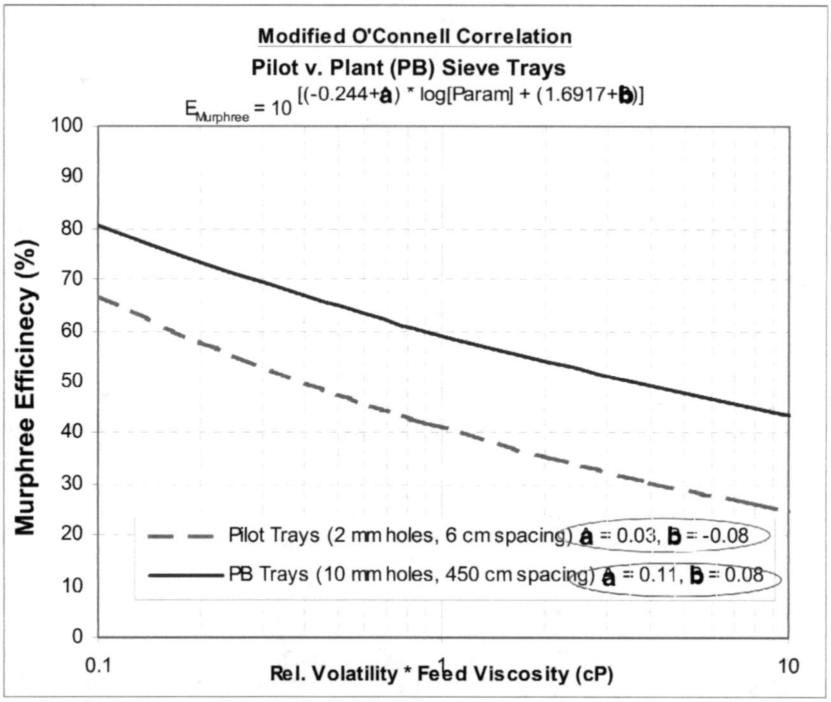

Figure 10

4. Design of New Column Using Efficiency Factors from an Existing Column

Based on the results from Example 2, the MOCA technique was then used to design a new full-scale side-draw column for the A-B-water ternary system, based on correction factors developed for an existing conventional plant column; the new column used the same tray geometry as the conventional column, albeit with a larger diameter. The resulting design required 1/5 fewer trays than was originally expected assuming efficiencies based on heuristics, resulting in project cost savings. The side-draw column design has been very successful – the separation performance almost perfectly matched the simulations with the MOCA-calculated tray efficiencies, and the operating column temperature profiles closely match those predicted by the simulations, as shown in **Figure 11.**

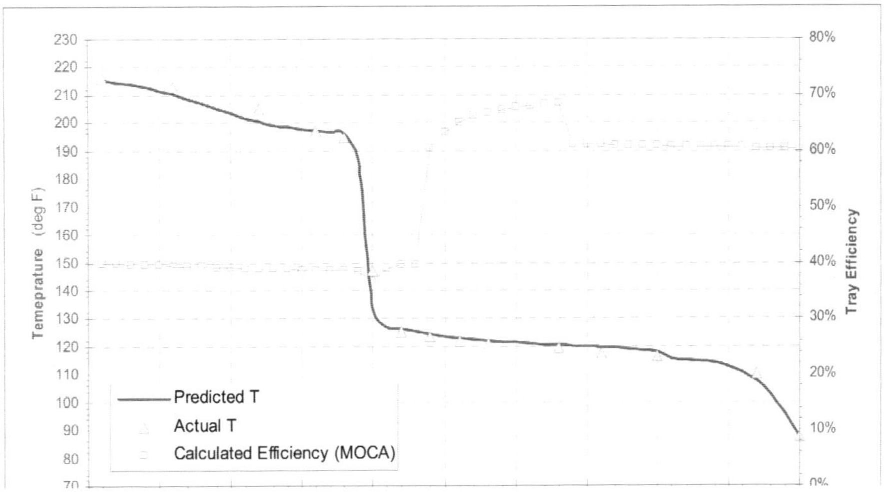

Figure 11

CONCLUSIONS

- The MOCA technique has a number of useful advantages for estimating tray efficiency, provided that both a trustworthy simulation and good data from existing units are available with which to determine the O'Connell correction factors: The technique is fast and easy use, taking advantage of modern software tools; it can be applied to columns where the relative volatility varies widely from top to bottom (e.g. azeotropic systems); and there is no need to obtain stage samples / compositions to calculate point efficiencies – temperature profiles are all that is needed.

- Efficiency transition zones do not necessarily coincide with changes from stripping to rectification. They appear to be primarily related to shifts in composition, and the accompanying changes in viscosity and relative volatility.

- MOCA technique has potential to yield more cost-effective column designs, by extending the O'Connell method to modern tray geometries. When known geometries are used, new units or retrofits can be specified with higher confidence in efficiency predictions, resulting in less conservative and more cost-effective designs.

- For existing columns, simulations based on theoretical stages and overall efficiencies can be revised using the actual number of trays and the calculated tray-by-tray efficiencies, allowing optimized operation, e.g. feed tray location, minimizing reflux, etc. It can also be used to identify operating problems, e.g. during periods when significant discrepancies in the actual vs. predicted temperature profiles occur.

FUTURE WORK

The broader value of the MOCA technique lies in the determination of (linearized) modified O'Connell plot slope/intercept (A,B) values, for the wide variety of tray styles and geometries used industry. While being an empirical approach, it would nonetheless become a useful tool for predicting the efficiencies of trayed columns, enabling more cost effective designs and more efficient operation of columns throughout the chemical and refining industries. To get to this point, the following work would need to be done:

- The general applicability of the MOCA technique needs to be validated. This includes confirming that the (a,b) correction factors for a given tray geometry are valid over a wide range of systems (e.g. pressures, component mixture types, sensitivity to foaming tendencies, etc.), as well as determining whether the MOCA technique can be extended to other types of trayed mass transfer operations (e.g. absorption, gas stripping, extraction).

- Developing a database of tray-specific (a,b) correction factors for a variety of tray styles and specific geometries. It would be interesting to determine whether there are common values or groupings of correction factors for "families" of tray types, e.g. sieve vs. valve vs. bubble caps vs. dual-flow tray, rectangular vs. circular holes, single vs. multi-pass, various hole cross-sectional areas, etc.

- Statistical analyses of the tray-specific (a,b) correction factors. The variability of the correction factors for given tray geometry (used in a variety of services) needs to be determined, to establish the confidence range for efficiencies predicted using MOCA technique.

- Determine whether the (a,b) correction factors or the corresponding (A,B) linearized modified O'Connell plot slope/intercept values correlate with tray design parameters (e.g. path length, bubbling circumferential area : hole cross section, tray spacing, % hole area, etc.).

(48e) On Distillation Tray Entrainment

M.R. Resetarits[1], T.J. Cai[1] and James R. Whiteley[2], (1)Fractionation Research, Inc., Stillwater, OK, (2) Chemical Engineering, Oklahoma State University, Stillwater, OK

Liquid tray-to-tray entrainment is not the sole precursor to tray flooding. The ability to predict entrainment does not necessarily assure the accurate prediction of flood points. Similarly, entrainment is not the sole reason for the efficiency deteriorations that sometimes occur at increasing rates and that usually occur near flood points. Jetting, spraying and contact times also impact those efficiencies detrimentally. The entrainment variable may have been over-used and over-quoted over the last 50 years. Recent experiences with closely-spaced trays have led investigators down other paths.

See more of this Session: Trays Hydraulic Regimes, Entrainment and Efficiency: Building on Dr. Fair's Foundations
See more of this Group/Topical: Topical 8: The Dr. James Fair Heritage Distillation Symposium

(48f) Modeling Fraction Jetting on Distillation Trays

Anand N. Vennavelli[1], James R. Whiteley[1] and M.R. Resetarits[2], (1)Chemical Engineering, Oklahoma State University, Stillwater, OK, (2)Fractionation Research, Inc., Stillwater, OK

The bi-phases on distillation trays can function in several different regimes depending on the physical properties of the system and the liquid-to-vapor ratio. Predicting the fraction of the vapor transported as jets, or fraction jetting, on a tray operating in the mixed-froth regime can bridge froth and spray regime models, explain the gradual changes in tray efficiency during the froth-spray transition, and eliminate the need to predict the froth-spray transition point when separate froth and spray regime models are used. Fraction jetting models will also facilitate multi-regime models, such as the Syeda, Afacan, and Chuang (2007) sieve tray efficiency model, that are valid for both froth and spray regimes. Current fraction jetting models are empirical and developed from limited data. In this paper, the factors affecting fraction jetting on a tray are discussed and a new semi-empirical model to predict fraction jetting as an alternative to existing models is presented.

See more of this Session: Trays Hydraulic Regimes, Entrainment and Efficiency: Building on Dr. Fair's Foundations
See more of this Group/Topical: Topical 8: The Dr. James Fair Heritage Distillation Symposium

(26d) Don't Let Baffle Tray Flood Baffle You

Henry Z. Kister and Matthew Olsson, Fluor Corporation, Aliso Viejo, CA

Understanding of baffle tray flooding.

See more of this Session: Trays Hydraulic Regimes, Entrainment and Efficiency: Building on Dr. Fair's Foundations
See more of this Group/Topical: Topical 8: The Dr. James Fair Heritage Distillation Symposium

Multi-pass distillation tray optimization by computational fluid dynamics and computational mass transfer method

ZHAO Dan , GAO Xueying , LIU Chunjiang , Yuan Xigang

([1]*State Key Laboratory of Chemical Engineering* , *Tianjin University* , *Tianjin* 300072 ,

Tianjin , *China*)

Abstract: A two-dimensional pseudo single phase computational fluid dynamics (CFD) and computational mass transfer (CMT) model have been presented to describe the flow and mass transfer behavior on commercial scale multi-pass tray. In this model, both the resistance and bubbling effect of the uprising gas phase are considered. Based on the proposed model, one case study was conducted by calculating the velocity profile and concentration regime on one four-pass tray with 12.6 m in diameter designed by equal bubbling area method. Simulated results show that the strong non-uniformity in flow pattern on side tray unfavorably influences the tray separation efficiency. And it also increases the sensitivity of tray efficiency as to different operational conditions. To optimize the tray configuration and thereby enhance the tray separation performance, modified tray structures were developed, which increase the Murphree plate efficiency by 4.53%~9.22%. Also for each tray configuration, the flow and mass transfer phenomenon were studied at different liquid loading ratios (Q_1/Q_2s) to obtain the optimal operation conditions. Finally potential reasons leading to the enhanced efficiency were analyzed.

Key words : multi-pass distillation tray; computational fluid dynamics; computational mass transfer ; tray efficiency ; flow pattern optimization

1. Introduction

Commercial scale multi-pass tray column has been widely used in vapor-liquid mass transfer process, especially in case of high liquid flow rates, such as high reflux ratio distillation and extraction distillation[1]. In industrial distillation operations, as the increase of liquid flow rates, the tray configuration is split into multiple streams or passes to accommodate high liquid loads[2]. Generally speaking, four-pass trays are recommended for services with high liquid loads in towers larger than 3.6 m in diameter[3].

Many studies have reported that the flow behavior of liquid on the tray deck has great impacts on the tray efficiency, especially in case of large-diameter columns[4-6]. In other words, uniform flow field will realize the maximized tray efficiency and irregular flow patterns will reduce tray separation capacity. Unfortunately, as the diameter and the liquid flow rates increase, the non-uniformity of fluid field becomes even more substantial, which has been validated by several literatures. Porter et al.[7] found in their work that there exists a stagnant zone in the arc-shaped region of large-diameter trays. Bell et al.[8] investigated the flow non-uniformities on commercial scale column trays by means of fibre optics technique. In their work, local recirculation phenomenon in the tray arc-shaped region was reported. Yu et al.'s[9] hydraulic experiment results indicated that, under certain operational conditions, recirculation zone also turned up at the two ends of the inlet weir besides of the arc-shaped region. Moreover, Yu et al.[10] also studied the flow patterns on two-pass distillation tray by measuring the residence time distribution on it. Their work revealed that, in two-pass distillation column, there exists recirculation area on the tray below the side down-comer while it does not show up on the tray below the center down-comer. However, up to now there are few papers focusing on the flow behaviors on the four-pass column tray. To accurately predict the separation efficiency on the four-pass distillation tray, detailed information about the fluid flow behavior on the tray is urgently required.

On the other hand, due to the large diameters of four-pass distillation tray, it is actually difficult and costly to make investigation by experimental methods. Fortunately, previous investigations well proved that the CFD-based methods could provide relatively precise prediction of the flow and mass transfer process on commercial scale distillation tray. By using the commercial software PHEONICS, Liu et al.[11] simulated the three dimensional velocity distribution on one-pass and two-pass distillation trays, which was well in accord with the relevant experimental results. Liu et al.'s[12] work insisted that it was really feasible to accurately calculate the velocity regime, concentration distribution and the tray efficiency by means of CMT methods on large-scale column trays. However, traditional multi-pass tray designs are mainly based on industrial experience and empirical models, which exhibits lots of limitations in accuracy of prediction. Up to present, there was very few published work on how to design multi-pass distillation tray by CFD method. In this paper, a new CFD model for its simple form with sufficient accuracy together with mass transfer equations have been utilized to describe the flow and mass transfer conditions on a four-pass distillation tray with 12.6 m in diameter. Using this new model, improved tray configuration and corresponding optimal operation conditions have been developed.

2. Physical Model and Computational Parameters

2.1 Traditional Designed Four-Pass Column Tray

In the traditionally designed four-pass distillation tray column, there are two different kinds of trays-"a-type" and "b-type", which are alternately installed inside the four-pass distillation column. On a-type tray, liquid flows from the side down-comer and the center down-comer to the off-center down-comers (OCDCs)[2]; on b-type tray, liquid flows from the OCDCs to the side down-comer and center down-comer, respectively. For the reason that there is no strong non-uniform flow behaviors, such as the recirculated flow, on the b-type tray indicated from the simulation results, the research focus in this work is mainly on the optimization of the a-type tray.

Generally speaking, the traditional design methods for four-pass column tray include equal bubbling area method and equal flow-path length method. The former one, employed in this paper, demand that the geometrical areas of both side and center tray are equal.

Considering the symmetrical nature of the round-shaped tray, only the top-left quarter of the whole tray, was calculated to reduce the computing cost. Figure 1 exhibits the final computational domain in this simulation, i.e. the computational physical model. In this figure, □ and □ denote the side tray and center tray, respectively. Additionally, the area above the dashed line is nominated as the arc-shaped region; the area below is referred as main-stream region.

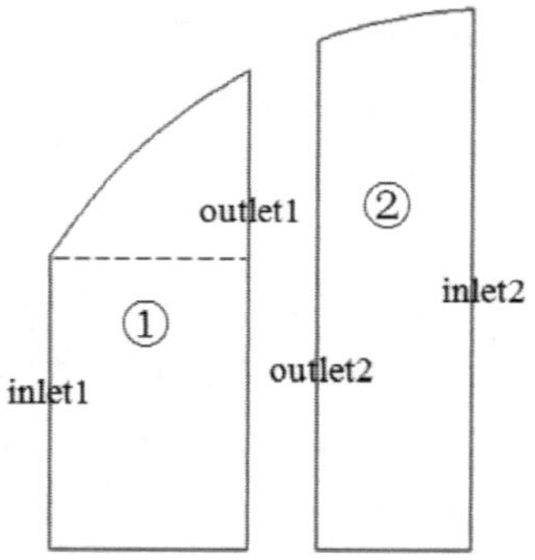

Fig. 1 physical model in this paper

Table 1 lists the relevant computational parameters involved in this simulation. Based on the assumption that the total liquid loading on the test tray is constantly kept as 0.27 $m^3 \cdot s^{-1}$, the corresponding liquid flow rates on side or center tray can be calculated according to different Q_1/Q_2s.

Table 1 Computational parameters

Liquid loading Q /(m^3/s)	Weir length l_1/m	Weir length l_2/m	Gas superficial velocity u_s/(m/s)	Initial liquid concentration x_0	Initial gas concentration y_0	Point efficiency E_{OG}	Relative volatility α
0.27	6.84	12.58	1.15	0.823	0.794	0.7	1.25

2.2 Modified Four-Pass Column Tray (□)

In the modified four-pass column tray (□), eight baffles are introduced to steer the fluid flow on the side tray without any changes on the center tray. Each baffle is approximately equal to one fourth of its corresponding extension line between the inlet and outlet of side tray. On the other hand, the side tray (top-left quarter) has been divided into five regions with the same geometrical area by the extension lines of the added baffles. Besides, the inlet weir of side tray has been divided into five equal-length parts by these baffles. The computational parameters for this new tray configuration are also shown in table 1.

2.3 Modified Four-Pass Column Tray (□)

To further optimize the tray configuration on the basis of the modified tray (□) and thereby increase the Murphree tray efficiency, the uppermost baffle in tray (□) has been extended to half of its extension line, referred as modified four-pass tray (□). Meanwhile, other baffles have been kept the same as those in tray (□).

3. Mathematical Model and Numerical Algorithm

In most cases of multi-pass tray operations, the large liquid flow rates within the column will induce large clear liquid height on the distillation tray, which is far beyond 2 cm. Therefore, the flow behavior on one single tray can be described by a two-dimensional model[13]. Besides, the new model in this work is considered as a quasi-single phase one, with both flowing liquid and holdup bubbles behaving as a "single fluid". Meanwhile, the effects of uprising gas phase on the fluid momentum and turbulent energy were taken into consideration by introducing source terms in transport equations. Furthermore, the k-ε equations were included in this model to close the modified Navier-Stokes equations.

4. Results and Discussions

4.1 Validation of the CFD model

As a result of the giant diameter of the test tray in this work, experimental validation seems to be unpractical. Fortunately, there are still two sorts of evidence to justify our proposed model. On the one hand, the recirculation region on the side tray is successfully predicted by the present model, which is well accord with relevant industrial observation. On the other hand, according to our simulation results, the recirculation region expands with the increase of Q_1/Q_2 for traditionally designed four-pass tray. And this trend of variance agrees reasonably well with the experimental results from Porter et al[14].

4.2 Traditionally designed tray configuration

First of all, the velocity distribution in the liquid phase reveals remarkable difference of flow patterns between the arc-shaped region and main-stream region: liquid flows much more uniformly in the latter region compared with that of the former one. Quantitatively speaking, the average velocity within the arc-shaped region is estimated to be 0.072m/s while that beyond this region is about 0.179m/s. Besides, there exists recirculation phenomenon on the tested side tray, which is generally supposed to badly impact the tray separation efficiency. As a result, the strong flow non-uniformity on the side tray deeply limits the tray separation capacity, indicating large room and great necessity for tray optimization. Furthermore, the concentration regime of light component exhibits high similarities with corresponding velocity distribution on the same tray, as shown in figure 2, which tells the importance of flow patterns as to the tray efficiency.

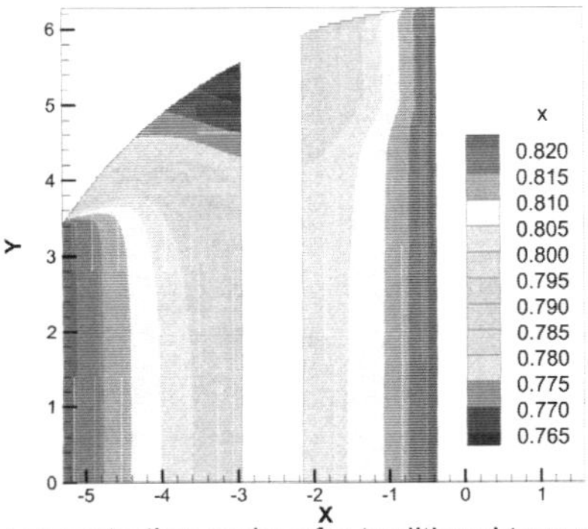

Fig.2 liquid concentration regime for traditional tray configuration

On the other hand, Murphree tray efficiency can be calculated as follows:

First, $\overline{x_{out}}$ (the average liquid concentration at the tray outlet) can be obtained from the concentration regime shown in figure 10. Following that, based on the assumption of constant relative volatility, $\overline{y_{out}^*}$ (the average vapor concentration in equilibrium with the outlet liquid concentration) is calculated from

$$\overline{y_{out}^*} = \alpha \overline{x_{out}} / \left[\alpha \overline{x_{out}} + \left(1 - \overline{x_{out}}\right) \right]$$

Meanwhile, $\overline{y_{out}}$ (the average concentration of vapor leaving the test tray) can be easily estimated on the basis that Murphree point efficiency within the whole computational domain is assumed to be the same. In the end, Murphree plate efficiency for the test tray can be given by

$$\mathrm{E}_{MV} = \frac{\overline{y_{out}} - \overline{y_{in}}}{\overline{y_{out}^*} - \overline{y_{in}}}$$

According to the procedures mentioned above, the Murphree plate efficiency for the side tray, center tray and the whole tray are predicted to be 0.825, 0.887 and 0.867, respectively. As expected, the tray efficiency on the side tray is much lower than that of the center tray due to its badly-distributed flow pattern. In other words, although the geometrical area between the side and center tray is actually equal, the effective tray area for vapor-liquid mass transfer on the two trays could be somewhat different resulted from the reasons discussed above.

Up to now, the widely-recognized problem in respect of four-pass tray design and operation is how to well distribute the liquid on different tray panels[15], i.e. the side tray and center tray in this paper. In fact, it is next to impossibility to ensure the distribution ratio to be kept unchanged during the whole operational process. As a result, it means a lot to investigate the effects of unstable liquid distribution between the side and center tray on the tray performance.

Figure 3 shows the predicted Murphree tray efficiency for the same tray under different Q_1/Q_2s ranging from 0.6 to 1.4. Firstly, there exists an optimal Q_1/Q_2 turning up around 0.8 in terms of tray efficiency. However, in the case of equal bubbling area method, the maximized tray efficiency is expected to be reached when the Q_1/Q_2 is at 1.0[2]. The shift of optimal Q_1/Q_2 from 1.0 to 0.8 should be subjected to the decrease of effective mass transfer area on side tray, making its separation capacity be largely reduced. Besides, the Murphree tray efficiency falls as the Q_1/Q_2 becomes too small or too large, probably resulting from the occurrence of localized flooding when excessive liquid flows to one part of the test tray[2]. Moreover, the tray efficiency decreases more steeply at low liquid loading ratios. For example, the efficiency discrepancy of Q_1/Q_2 at 0.7 relative to Q_1/Q_2 at 1.0 is 0.0237 while that is only 0.0045 in case of Q_1/Q_2 at 1.3. In a word, this strongly asymmetrical trend of efficiency variance increases the instability of four-pass column performance, and hence poses great risks to the goal of keeping constant concentration in the distillate stream, all of which enhance our desire to improve the traditional four-pass tray configuration.

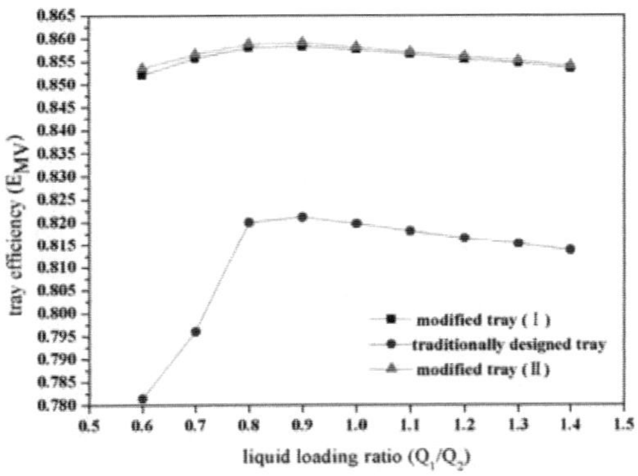

Fig. 3 Murphree tray efficiency-liquid loading ratio curve for different tray configurations

4.3 Improved tray configuration

As indicated from the simulation results for the modified tray (□), it is obvious that the flow pattern on the side tray becomes much more regular with the disappearance of recirculation region. Quantitatively speaking, the average velocity within the arc-shaped region for this new tray is estimated to be 0.170m/s and that beyond this region is about 0.162m/s. Similarly, it can be seen from figure 4 that the concentration distribution also gets much more uniform. As a result, the calculated tray efficiency for the modified tray (□) is increased to 0.908 with a percentage of 4.7% compared with that of the traditionally designed tray, indicating that the introduction of baffles on the side tray can improve the flow patterns on it and thereby enhance the tray performance.

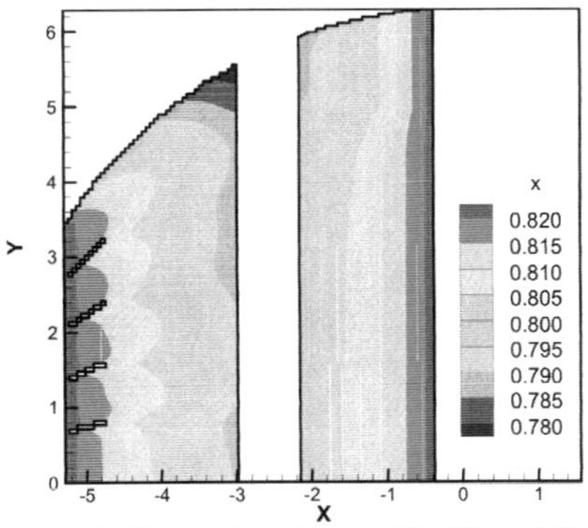

Fig.4 liquid concentration regime for modified tray (□) configuration

What's more, it can be seen from figure 3 that, due to the forced flow guidance of added baffles, the Murphree tray efficiency at each Q_1/Q_2 gets remarkable increase. And it increases much more largely at lower liquid loading ratios, which fortunately reduces the influence of Q_1/Q_2 on the tray efficiency to be less sensitive. As a result, by introducing certain baffles, the four-pass column tray could be ensured to operate in a relatively stable state with the Q_1/Q_2 varying within a wide range.

Figure 3 also shows the predicted tray efficiency under different liquid loading ratios for modified four-pass column tray (□). It reveals that change of the length or the angle of baffles can slightly change the tray performance. As a result, for real-world four-pass tray designs, one could get the optimal tray configuration through "try and error" method by taking consideration of both tray efficiency and pressure drop between adjacent trays.

4.4 Detailed reasons for efficiency improvement

It is generally recognized that there are two most important factors-the vapor-liquid mass transfer time and the flow pattern- that could strongly influence the distillation column performance. On the one hand, the liquid residence time on the tray is mainly determined by the flow velocity and the flow-path length. As to the case study in this paper, both inlet velocity and flow-path in the liquid phase are almost kept unchanged or just a little changed. As a result, the vapor-liquid contacting time could not be the main reason leading to the enhanced tray efficiency. Another evidence to support this conclusion is that the modified tray efficiency at lower Q_1/Q_2s with longer liquid residence time unexpectedly gains larger increase than that at higher ones with shorter vapor-liquid contacting time. However, when the Q_1/Q_2 is larger than 0.9, it is probable that the increase of mass transfer time that induces higher increase in tray efficiency at higher Q_1/Q_2s.

Up to now, it can be concluded that the main factor controlling the four-pass distillation tray performance should be the flow patterns. It has been reported that the presence of non-uniformities including recirculated flow, stagnant flow and so on, in the liquid as it flows from inlet to outlet that could largely reduce the tray performance. In this paper, only liquid recirculation and velocity distribution have been studied. In case of one-pass distillation column, the recirculation existence can be a major cause of the tray efficiency decrease according to the reports from Liu et al. However, in respect of four-pass distillation tray, the recirculation percentage is relatively small, which is only around 0.13% of the total side tray area. Furthermore, how much influence the existence of recirculation region has on the tray performance is also investigated in this work. To that end, a new tray configuration is developed by keeping the uppermost baffle with removal of the other three baffles below. When the length of the added baffle is gradually increased from zero, the tray efficiency is expected to increase with the decrease of the recirculation area. According to our simulation results, it is pretty easy to eliminate this recirculation with little change of the flow pattern on the side tray. Besides, although the recirculation region accounts for only a small percentage of the total tray area, it can contribute a lot in terms of the deterioration of tray performance. For example, in case of Q_1/Q_2 at 1.0, when the baffle length is only 0.156m, the recirculation phenomenon can be totally prevented, and meanwhile the tray efficiency is increased by 1.24%. Considering that the tray efficiency increase in modified tray configuration (□) at the Q_1/Q_2 of 1.0 is about 4.62%, elimination of the relatively small proportion of recirculation makes up 26.91% of the total tray efficiency increase. And the other more than 70% should be owing to the improvement of flow patterns on the side tray. In real-world four-pass distillation tray design, although the recirculation phenomenon can bring much change in tray efficiency, it can be removed with little difficulty. As a result, our efforts are supposed to focus on the optimization of the flow pattern.

5. Conclusions

A CFD model has been proposed in this paper to describe the flow and mass transfer phenomenon on four-pass distillation tray. Despite lack of experimental invalidation, this new model has been justified to a certain extent by comparing the simulation results with those reported in previous papers and industrial observation. Then with this newly presented model, a case study is conducted on one four-pass distillation tray with 12.6m in diameter designed by equal bubbling area method. Calculated results reveal that both the strong non-uniformities on the side tray and the intense sensitivity of tray efficiency as to operational conditions pose the four-pass distillation tray in an unsatisfied and unreliable performance. As a result, modified tray configurations are developed to enhance the tray separation capacity. And simulated results exhibit that the Murphree tray efficiency has been largely increased compared with that of traditionally designed one at different liquid loading ratios. Also the stable trend of variance in tray efficiency for the modified trays shows great promise to ensure the performance of the whole distillation column even if its internal vapor-liquid distribution is actually not quite well. Finally, in order to provide some instructions in the real-world four-pass column tray design and optimization, detailed reasons leading to the enhanced tray efficiency for modified trays have been analyzed. First, the vapor-liquid residence time on the test tray may contribute a lot to the increased tray efficiency when the Q_1/Q_2 is larger than 0.9 while it can be neglected at low Q_1/Q_2s. Second, the existence of recirculation region on the side tray actually produces a large decrease in tray efficiency. Fortunately, since it is really small and easy to be removed, not too much attention need to be paid during four-pass tray design and optimization. Instead, the flow patterns on the four-pass tray are supposed to play the most important role in determining the column tray performance.

References

[1] Sun Lanyi, Chen Liang, Guo Huixiang, Li Jun, Li Qingsong. Design of multi-pass trays. *Modern Chemical Industry*, 2008, **28**(1): 25-28

[2] Mark P. Ensure proper design and operation of multi-pass trays. *Chemical Engineering Progress*, 2005, **101**(6): 22-27

[3] Henry Z K, Richard D, Walter J S, Matthew R O. Preventing Maldistribution in Multi-Pass Trays. *Chemical Engineering Progress*, 2010, **106**(4): 32-41

[4] Bell R L, Solari R B, Effect of nonuniform velocity fields and retrograde flow on distillation tray efficiency. *AICHE Journal*, 1974, **20**(4): 688-695

[5] Proctor, Sebastian J, Biddulph, Michael W, Krishnamurthy, Krish R. Effect of liquid channeling on a 1.8-m distillation sieve tray. *Industrial & Engineering Chemistry Research,* 1998, **37**(6): 2535-2540

[6] Porter K E, Davies B, Enjugu B A. Investigation of the effect of the liquid flow pattern on sieve tray performance by means of the water cooling technique. *IchemE. Symp. Series,* 1987, **104**: A569-A588

[7] Porter K E, Lockett M J, Lim C T. Effect of liquid channeling on distillation plate efficiency. *Trans. Inst. Chem. Engrs.,* 1972, **50**(2): 91-101

[8] Bell R L, Residence time and fluid mixing on commercial scale sieve trays. *AICHE Journal*, 1972, **18**(3): 498-505

[9] Yu Guocong, Huang Jie. Simulation and efficiency of large tray (□)—eddy diffusion model for non-uniform velocity field. *Journal of Chemical Industry and Engineering* (*China*), 1981, **32**(1): 11-21

[10] Yu Guocong, Huang Jie, Zhang Zeting. Residence time profile and plate efficiency for a large tray with single-pass or two-pass liquid flow. *Journal of Chemical Industry and Engineering* (*China*), 1986, **2**: 151-161

[11] Liu B T, Liu C J. Three dimentional simulation of liquid flow on distillation tray. *Chinese Journal of Chemical Engineering*, 2002, **10**(5): 517-521

[12] Liu Chunjiang, Cheng Hong, Yuan Xigang, Yu Guocong. Simulation of the velocity field on distillation tray with consideration of gas phase effect. *Journal of Tianjin University*, 2001, **34**(6): 741-744

[13] Liu Chunjiang, Yuan Xigang, Yu Guocong, Zhu Xiaojing. Simulation of the velocity field on a distillation tray with consideration of gas phase effect. *Journal of Chemical Industry and Engineering* (*China*), 1998, **49**(4): 483-488

[14] Porter K E, K T Yu, Chambers S. *In: Int Chem E Symp Series*. London: Hemisphere Publishing Corporation, 1992, **128**: A257

[15] Daniel R S. Preventing Designing four-pass trays. *Chemical Engineering Progress*, 2010, **106**(4): 26-31

(71a) Research on Mass Transfer Columns "Old Hat or Still Relevant"

Michael Schultes, Raschig Ring Department, RASCHIG GMBH, Ludwigshafen, Germany

The past few years have witnessed a major decline in university-based research in the area of mass transfer columns although up until only about 10 years ago scientific research and communication at the international level were still highly intensive.

In the United Kingdom, for instance, P.V. Danckwerts developed mass transfer models for the calculation of mass transfer processes with a superimposed chemical reaction, and K.E. Porter worked intensively in the field of fluid dynamics and mass transfer efficiency of trayed and packed columns. In Japan, K. Onda developed equations for the calculation of mass transfer coefficients, and numerous other Japanese departments worked on describing chemical reactions in mass transfer columns. In the USA, T.H. Chilton and P.A. Colburn developed mass transfer theories based on driving forces, T.K. Sherwood developed calculation approaches for mass transfer coefficients using dimensionless numbers, E.V. Murphree elaborated degrees of plate efficiency, W.K. Lewis developed analogous relationships between heat and mass transfer in two-phase counterflow columns and J.R. Fair created equations for mass transfer coefficients for modern tower internals. In Germany, E. Kirschbaum worked on the fundamental calculation methods for the design of distillation columns, and A. Mersmann, R. Billet and J. Stichlmair worked, for instance, on mass transfer correlations for modern internals. These are only a few of the departments of that era to make a name for themselves in the field of mass transfer technology in two-phase mass transfer columns.

Meanwhile it has got quiet concerning university-based mass transfer research with trayed or packed towers. Many departments have been closed down or reassigned to other fields, e.g. the chair for mass transfer engineering of Z. Olujic at the Delft University of Technology in the Netherlands. Why is this? The long economic recession in the chemical and related industries in the last decade of 20th century led to a drastic reduction in the numbers of students in this field, and the financial constraints of the public sector reached the universities. Departments needing large amounts of funds for research were abandoned, possibly also in the belief that the research results from the 20th century would be sufficient to capture the future scope of trayed and packed column mass transfer engineering.

Recently in Germany a discussion restarted about the requirement of research in the field of mass transfer columns. It was concluded that the prediction of fluiddynamic and mass transfer efficiencies in trayed and packed columns are still too vague to minimise the investment cost and energy consumption of mass transfer towers. To reach this target a better experimental data base must to be available which is why new standards have to be defined to constitute testing procedures.

The paper gives an overview about the status of the predictability of mass transfer columns and will show topics to be considered for future research in this area. The question to be discussed is whether this research task is **"Old Hat or Still Relevant"**.

See more of this Session: New Frontiers in Packing Technology: Continuing Dr. Fair's Great Work
See more of this Group/Topical: Topical 8: The Dr. James Fair Heritage Distillation Symposium

(71b) Predicting the Liquid-Side Mass-Transfer Resistance of Structured Packings: State of the Art

Zarko Olujic, Process & Energy Laboratory, Delft University of Technology, Delft, Netherlands and A. Frank Seibert, Separations Research Program, The University of Texas at Austin, Austin, TX

Well established correlations for estimating the liquid-side mass transfer coefficient of structured packings have been thoroughly evaluated using experimental evidence on overall performance obtained in total reflux distillation tests with various systems as employed at FRI, SRP, TUD, Sulzer, Bayer TS, etc. Strikingly large differences are found between different methods with respect to both the absolute values of mass transfer coefficients and the trends with respect to increasing vapor and liquid loads. This and other issues particularly that about the magnitude of relative contribution of liquid-side to overall mass transfer resistance, which according to new data seems to be larger in distillation than generally believed, are discussed and aview on the way forward including research needs and suggestions for implementing physically sound improvements to existing predictive models given.

See more of this Session: New Frontiers in Packing Technology: Continuing Dr. Fair's Great Work
See more of this Group/Topical: Topical 8: The Dr. James Fair Heritage Distillation Symposium

(71c) INTALOX ULTRA Random Packing - Pushing the Envelope

Izak Nieuwoudt, Koch-Glitsch, Wichita, KS and Pat Quotson, Koch-Glitsch, LP, Wichita, KS

Random packing is commonly used in absorption or light hydrocarbon distillation applications since the high liquid rates and/or high vapor densities complicate the use of other mass transfer devices. The capital and operating costs of such units can be reduced by using random packing with low pressure drop, high capacity and good separation efficiency. INTALOX® ULTRA™ random packing is a recent development that meets these criteria for improved performance. Data generated over a wide range of operating conditions will be presented in this paper. This include data from Koch-Glitsch's pilot plants and from F.R.I. Industrial case studies will also be covered.

See more of this Session: New Frontiers in Packing Technology: Continuing Dr. Fair's Great Work
See more of this Group/Topical: Topical 8: The Dr. James Fair Heritage Distillation Symposium

(71d) Transport Characteristics of Modern Structured Packing Types for Absorption and Distillation Columns (Mellapak250Y, 350Y, 452Y, 500Y)

Frantisek J. Rejl[1], Lukas Valenz[2] and Vaclav Linek[2], (1)Department of Chemical Engineering, Institute of Chemical Technology, Prague (ICT), Prague, Czech Republic, (2)Chemical Engineering, Institute of Chemical Technology, Prague (ICT), Prague, Czech Republic

The work presents absorption transfer characteristics for modern and industrially often used structured packing types Mellapak250Y, 350Y, 452Y a 500Y. The characteristics were determined using the standard measuring methods (Rejl et. al., 2009). *Liquid side volumetric mass transfer coefficient,* $k_L a$, *were* determined using an absorption system O_2/air ñ water, in which the main mass transfer resistance is on the liquid side. Oxygen desorption from water into a stream of nitrogen were used (Linek et al., 1984). The $k_L a$ values were calculated from the readings of oxygen probes in liquid samples obtained from the ëhead' and ëbottom' of a packing. *Gas side volumetric mass transfer coefficient,* $k_G a$, were determined using an absorption system SO_2 (0.03 vol. %) /air - 1 M NaOH solution. In this absorption system, the main resistance against mass transfer is on the gas side. For *effective interfacial area* measurement, chemical method using absorption of diluted CO_2 from air (1 vol. %) into a NaOH solution of initial concentration 1 kmol/m^3 were used. Characteristics were obtained in the flow range commonly used in the course of absorption operations, and of vacuum and atmospheric distillation operations. The $k_G a$ data for mentioned structured packing types have not been till this time published.

See more of this Session: New Frontiers in Packing Technology: Continuing Dr. Fair's Great Work
See more of this Group/Topical: Topical 8: The Dr. James Fair Heritage Distillation Symposium

(71e) Experiences with the Testing of Liquid Distributors

German A. Luna Sr., HSE, ACM FACILITY SAFETY, Calgary, AB, Canada

Spray nozzle & gravity distributors are primarily used in scrubbers and in wash zones, condensing zones, and pump around zones of crude oil vacuum towers. In many cases testing is limited due to time and equipment involved; this normally creates a conflict between maintenance and Process, and many times the test is not done. This paper presents the importance of the test, experiences performing the test and a "step by step" guideline for the performance of the test from the initial preparation until the end. Safety guidelines are also provided as a part of the "step by step" procedure.

See more of this Session: New Frontiers in Packing Technology: Continuing Dr. Fair's Great Work
See more of this Group/Topical: Topical 8: The Dr. James Fair Heritage Distillation Symposium

(71f) Hydrodynamic Study of KATAPAK-SP11 Structured Packing with Multiphase CFD and Heat Integration

Jing Huang, Robert W. Lyczkowski, Chandrakant B. Panchal and Richard D. Doctor, Energy System Division, Argonne National Laboratory, Argonne, IL

Hydrodynamic Study of KATAPAK-SP11 Structured Packing with Multiphase CFD and Heat-integration

Jing Huang, Robert W. Lyczkowski, Chandrakant B. Panchal, and Richard D. Doctor

Argonne National Laboratory; 9700 S. Cass Avenue; Argonne, IL 60439-4815 USA

Abstract

The number of actual reactive distillation installations is quite small despite a large number of patents and extensive academic and industrial research. The main reason for this is that it is difficult to match the optimum conditions for distillation, reaction, and scale up of experimental test data to commercial units [1]. Pressure is the primary variable that affects efficient separation. Many reactive distillation processes are not very flexible due to the interactive effects of reaction and separation on each reactive stage. Single- and two-phase flow through a 100 mm diameter Katapak-SP11 structured packing [2] was modeled using ANSYS FLUENT 12.1 [3]. The purpose of this analysis is to determine the liquid flow distributions interior and exterior to the catalyst baskets, catalyst wetting, and hence reaction effectiveness. The analysis is expected to help optimize catalyst loading on individual structured catalytic packings, as well as for the entire column. The coupled catalyst baskets and corrugated flow channels were treated using the new dual-porous media formulation. Because of the complex structure of the corrugated flow channels, they were treated as one porous medium and the catalyst baskets containing, approximately 1 mm diameter catalyst particles, were treated as another. Different values of porosity and permeability were used for each porous medium region. This new formulation vastly simplifies modeling of complex structured packings performed in the past [4] in order to understand their hydrodynamics. This novel modeling approach will allow optimum heat-integration. Calculations for dry and wetted packings (only gas or liquid flow though the column) were carried out. The dry packing pressure drop was compared with available data. Subsequently, two-phase, counter-current liquid-gas flow through the structured packing column was modeled. A range of gas and liquid flow rates was investigated to determine pressure drop and liquid flow distributions in the catalyst baskets and flow channels and wetting of the catalyst baskets.

References

1. Harmsen, G. Jan, "Reactive distillation: The front-runner of industrial process intensification: A full review of commercial applications, research, scale-up, design and operation," *Chemical Engineering and Processing* 46 (2007) 774ñ780

2. Aferka, S. *et al., Liquid Load Point Determination in a Reactive Distillation Packing by X-Ray Tomography.* Canadian Journal of Chemical Engineering, Vol 88, pp. 611-617 (2011).

3. ANSYS FLUENT 12.1 User's Guide, ANSYS, Inc., Canonsburg, PA (2009).

4. Vervlot, D., et al., *Intensification of Co-current Gas-Liquid reactors Using Structured Catalytic Packings: A Multiscale Approach.* Catalysis Today. Vol. 1475, pp. 5138-5143 (2009).

See more of this Session: New Frontiers in Packing Technology: Continuing Dr. Fair's Great Work
See more of this Group/Topical: Topical 8: The Dr. James Fair Heritage Distillation Symposium

(71g) Low Energy Direct Contact Condenser Designs for Claus Tail Gas Treating Units In Desert Environments

Henry Z. Kister[1], Thomas K. Chow[2], Dick B. Nielsen[2] and Arnold E. Sanchez[2], (1)Fluor Corporation, Aliso Viejo, CA, (2)Energy and Chemicals, Fluor Corporation, Aliso Viejo, CA

In a conventional Claus/Hydrogenation Sulfur plant design, tail gas from the Claus plant is hydrogenated to convert all the sulfur species to H2S and the tail gas is then cooled in a Direct Contact Condenser (DCC) before the H2S is removed from the tail gas by treating with generic or formulated MDEA or a specialty solvent such as ExxonMobil's Flexsorb SE® or SE Plus®. The feed gas to the MDEA or Flexsorb system has to be cooled in the DCC to a certain minimum temperature in order to meet SOx emission requirements.

In many situations, especially in arid regions, DCC air coolers cannot cool sufficiently to meet SOx emission requirements and cooling water is usually not available for trim cooling. Instead of using cooling water for DCC trim cooling, many Claus/Hydrogenation Sulfur plants in these environments must use refrigerant to cool the tail gas to the appropriate temperature.. Conventional DCCs utilize a single pumparound that takes the process water from the bottom of the DCC column, through air coolers, followed by cooling water or refrigerant coolers for trim cooling, and then back to the top of the column. For this type of design, the air coolers provide about half the duty and the refrigerant or water coolers provide the other half.

Fluor has invented a new DCC design that transfers as much as 50% of the trim cooling duty to the air coolers. An evaluation of a recent large-scale Middle East project indicated that this new design would result in about $20 MM in power savings over the 20-year life of the plant and an additional $70 MM in capital cost savings. Other Claus/Hydrogenation sulfur recovery facilities can achieve significant savings in proportion to the size of the plant.

This paper provides a technical background for Fluor's new patent pending Direct Contact Condenser technology, and discusses the economic advantages and operational flexibility that can be realized through its implementation.

See more of this Session: New Frontiers in Packing Technology: Continuing Dr. Fair's Great Work
See more of this Group/Topical: Topical 8: The Dr. James Fair Heritage Distillation Symposium

(71h) Transport Characteristics of Mellapak 250Y Packing Under Distillation Conditions. Effect of Axial Mixing of the Phases

Frantisek J. Rejl, Lukas Valenz and Vaclav Linek, Department of Chemical Engineering, Institute of Chemical Technology, Prague (ICT), Prague, Czech Republic

Transport characteristics (*HETP*, $k_L a$, $k_G a$) of metal Mellapak 250Y structured packing has been measured in the distillation column distilling three alcoholic systems (*methanol-ethanol*, *ethanol-propanol*, *methanol-propanol*) and one standard hydrocarbon system *(cyclohexane-heptane)* under total reflux condition at atmospheric pressure. Volumetric mass-transfer coefficients have been determined using "profile method".

Experimental aparatus was an atmospheric distillation column of an inner diameter of 0,15m packed with ten pieces of packing with the total height of the packing 2,1m. The column is equiped with seven sampling points along the packing enabling withdrawal of a sample of liquid and vapour directly from the packing and therefore measurement of concentration profiles in both phases. Experiments were performed for four powers into a reboiler up to u_G = 1,7m/s.

An axial mixing in terms of Bodenstein numbers Bo_L, Bo_G has been measured independently in the absorption column with water-air system. There has been found that axial mixing data transfer from absorption to distillation can be succesfully achieved using relation suggested by Fair (1999). Volumetric mass-transfer coefficients evaluated with axial mixing being taken into account differ substantially from those evaluated with common plug-flow assumption. Results indicate significant mass-transfer resistance in both phases and significant influence of axial mixing of the liquid phase on separation process. The effect of gas phase axial mixing was found negligible. Dependence of volumetric mass-transfer coefficients on the phase flow rates differs from the ones incorporated in commonly used models by Olujic', Billet-Schultes and Rocha-Bravo-Fair.

See more of this Session: New Frontiers in Packing Technology: Continuing Dr. Fair's Great Work
See more of this Group/Topical: Topical 8: The Dr. James Fair Heritage Distillation Symposium

(102a) Distillation Contributions of Dr. Jim Fair

M.R. Resetarits, Fractionation Research, Inc., Stillwater, OK, R. Bruce Eldridge, Chemical Engineering, The University of Texas at Austin, Austin, TX and A. Frank Seibert, Separations Research Program, The University of Texas at Austin, Austin, TX

Dr. Jim Fair's contributions to the distillation technology will not be forgotten. The books, handbook chapters, papers and articles that he authored and coauthored while he was at Monsanto and at the University of Texas at Austin formed, in essence, the basis of many of the things that we know today about trays, random packings and structured packings. He, his colleagues and his graduate students provided invaluable information regarding capacities, pressure drops, efficiencies, scale-up, dualflow, cocurrent flow, debottlenecking and direct-contact heat transfer. The models and correlations that were developed are still quoted globally regularly. Dr. Fair will be missed – globally, regularly.

See more of this Session: Contributions to Distillation Troubleshooting and Control
See more of this Group/Topical: Topical 8: The Dr. James Fair Heritage Distillation Symposium

(102b) Troubleshooting - Strategies and Tips From Those Who Do It Best

Ruth Sands, Heat, Mass & Momentum Transfer Group, DuPont Engineering Technology, Wilmington, DE

Two different troubleshooting strategies from the literature are compared, and practices and tips from expert troubleshooters are shared. Several examples illustrate these techniques. A particular case of a distillation tower experiencing poor efficiency is highlighted to demonstrate some solid fundamental methods, and it will also reveal some unexpected results.

See more of this Session: Contributions to Distillation Troubleshooting and Control
See more of this Group/Topical: Topical 8: The Dr. James Fair Heritage Distillation Symposium

Investigation of foam in distillation and absorption columns

Senger, G.; Wozny, G.: TU Berlin, Chair of Process Dynamics and Operation, Strasse des 17. Juni 135 Sekr. KWT9, 10623 Berlin / Germany

A physically accurate, analytical approach of foam is currently not possible. Due to its complexity, the topic of foam in mass transfer columns has not been adequately studied yet and provides an extensive research demand. As part of a running research project, the fluid dynamic performance (pressure drop, holdup) of different packings are determined for foaming systems. In addition, the location within the column where foam occurs is determined separately. In a further step, a new test cell is developed which characterizes the inclination of a foam material system with respect to the use in packed columns. Results of the running research project with regard to the phenology of the foaming behavior, the performance data in column operation with foaming media and the attempts of the developed test cell are discussed.

Introduction

Foam in the process industry can cause severe problems in distillation and absorption columns, which may result in increased pressure loss. Therefore, reduced capacity, reduced separation efficiency and the contamination of products in associated process units can be observed. If the process is no longer operable, the complete shutdown causes a corresponding loss of production. Foam problems often cannot or only insufficiently be predicted and prevented before. In addition, foam as the respective cause is often not recognized. In a field study of 31 different failures in mass transfer columns, foam related operating problems ranked with 51 of over 900 cases in 11th place [1].

Foam is a phenomenon which is highly dependent on the physical properties of the system and can occur more or less intense. Mass transfer columns are affected by foam in processes as crude oil, alcohol, organic acid and biodiesel distillation, extractive distillation, amine and sour gas scrubbing (e.g. CO_2 absorption), especially in the presence of liquid hydrocarbons [1], [2], [3]. The presence of fine particulate matter can promote foaming as well [4]. A detailed overview with references according to individual applications is given in [5].

Foam can be described by gas bubbles separated between liquid films. Conditional on a capillary force balance [6], three liquid films (lamellas) form a so-called plateau channel at an angle of 120° (Figure 1) [7]. The plateau channels are the connection between the foam lamellas and allow fluid exchanges within the foam structure. The presence of surfactants, which lower the surface tension, allow the gas bubbles flexibility as well as stability and therefore promotes foaming.

While some fluids, especially pure liquids do not foam at all [8] [9], other liquids foam more or less intense. The decay time of foams can range from a few seconds until several years - under undisturbed laboratory conditions. The explanation for such a different behavior can be expressed by thermodynamic and mechanical stability considerations, which are based on two fundamental mechanisms.

Due to gravity, the interlamellar fluid drains towards the lower foam layers and thins the upper foam structure out of liquid. When film thickness falls short of a critical laminar thickness respectively a mechanical force appears, this effect results in the destruction of the film layer. While the liquid drainage is destabilizing the gas bubbles and disintegrates the foam layers at the top, this liquid flow stabilizes the lower situated gas bubbles. In mass transfer columns, such a foam-stabilization by replenishing from above is caused by feed and reflux streams.

In addition, a pressure difference due to different curvature of the phase boundaries between the lamellae and plateau channels effects the foam destruction. According to Laplace's law, in the lower curvature of the phase boundary between liquid and gas in the lamella causes a higher pressure than in the plateau channel with its higher phase boundary layer curvature. This pressure difference accelerates the thinning of the stabilizing lamella liquid between the gas bubbles (Figure 1).

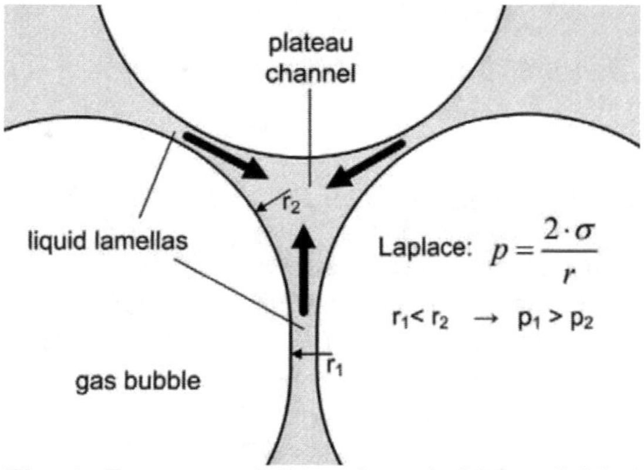

Fig. 1: Foam consisting of gas bubbles divided by liquid lamellas and plateau channel (p = pressure, r = curvature radius, σ = surface tension)

The destruction of foam bubbles is counteracted by the stabilization of the foam lamellae due to various effects. Foaming related operation problems in mass transfer columns are related primarily by the Marangoni effects and the so-called "Ross-type" foaming.
The Marangoni effects describe an elasticity increase of liquid films covered with surface-active substance.
Ross and Nishioka [10] have shown that the maximum foaming potential of various unmixing binary and ternary systems occurs in the single-phase region near the critical points. This is designated as so-called "Ross-type foam" [5], [11]. The appearance of a second phase acts as a foam inhibitor, so that in the two-phase region no stable foam occurs. For multi-component systems, they postulate a similar behavior.

In the literature there are several publications which deal in detail with foam stability and destruction [6], [8] [9], [12], [13]. Currently, the foam behavior can neither analytically nor thermodynamically be described accurately on the basis of physical property data. On the basis of existing hydrodynamic models there is also no reliable prediction of the occurrence and intensity of foam and its consideration for the process design yet possible.

Foam in mass transfer columns

In distillation and absorption columns, foam can occur in different places, such as in the evaporator, in the bottom (Figure 2), in liquid distributors or within the packing (Figure 3). In the latter case, there is the possibility that foam is discharged into the bottom or head. The recognition of a foam problem is limited since columns are difficult to observe on sight.

Since the foam potential in columns cannot be predicted yet, the existing industrial solutions for elimination or limitation of a foam problem are over-sizing or the search and usage of foam-suppressing substances.
Although the use of chemical foam inhibitors is undesirable for a variety of reasons such as additional costs and product contamination [1], [5], [14], their use is a common practice.
The foam inhibitor effect is based on the ability of a spontaneously spreading across the foam surface. This is only possible if the surface tension of the surface-active inhibitor is less than that of the foaming liquid [15]. Due to a fast propagation speed of the inhibitor on the phase interface, a several micron thick layer of liquid is carried away, which displaces the adsorption layer of the foam-stabilizing material [8] [9]. The lamellas thin out, although occupied by the surface-active inhibitor. Since the inhibitor is not enough stabilizing as the primary surfactant, the lamellas collapse due to their small thickness even at low mechanical stress.
Mechanical foam destruction mechanisms, which are used in bubble columns, are more advantageous compared to chemical methods, since products are not contaminated. In this case, the foam destruction is based on the effect of rapid pressure changes as well as on shock, compression and shear forces. The application of mechanical destruction methods on packed columns with internals is hardly transferable.

Fig. 2: Column bottom filled with foam (column Ø 300 mm, butanol-water / air, L= 50 m³/m²h)

Fig. 3: Foam in a structured packing (B1-350M, Ø 300 mm, L= 50 m³/m²h, F-factor 0.5 $Pa^{0.5}$, butanol-water / air)

A comparative analysis of different column internals (trays, centrifugal and structured packings) [16] comes to the conclusion, that the lowest capacity loss occurs in the case of structured packings and that they therefore are ideal for separation processes with foam problems. Although packed columns are under foaming conditions generally more suitable as tray columns, they can still be prone to foam. For the design of packed columns under foaming conditions there are no design advices published. For tray columns, however, there are empirical safety factors depending on the material system and the specific application published [11], but an improved pressure loss calculation is thereby not given.

While in the literature various fundamental studies on the foam stability and destruction can be found, beyond preliminary works carried out at the author's department [2], [17], [18], just a few systematic, scientific studies on the hydrodynamics of foaming media in packed columns were done [19]. In a running "Applied Industry Research Foundation" (AiF) project "Foaming media in packed columns" (AiF no. 16 073 N), extensive systematic studies on the foam behavior in columns are carried out.

Experimental foam tendency determination

Since physical properties such as surface tension, viscosity and density don't contain direct information on the foaming behavior, foaming material system are not yet analytically describable. For the experimental determination of the foam stability in applications in the food and surfactant industry, various methods like the Bikerman-index [20] or stability-index (for mass transfer induced foaming) are known. However, none of the methods claim to be universally applicable and be transferable to columns.

A rough classification of foam stability test methods results in a subdivision into static and dynamic methods; overviews can be found in Denkov [13] Prud'homme and Khan [21] and Bikerman [20].
The static methods for example produce foam by aerating a liquid sample or by pouring it from one container to another at a defined height. The foam stability evaluation is obtained by comparing either the generated foam volume, the deviation of the bubble size distribution or the time until the last bubble disappears. The dynamic methods often gain steady state conditions by aeration of a liquid sample - foam formation and destruction are here in a dynamic equilibrium. As a foam stability rating scale, the generated foam height or volume are taken in account, which result in characteristic parameters as specific times, foam volumes or relationships.

Desirable for the classification of foam stability is the determination of a characteristic parameter based on physical characteristics. The Bikerman foam generation procedure, also called foam-rise method, is carried out by aerating a liquid sample with a gas flow through a porous glass plate (frit) or a pack of capillary tubes. The use of nitrogen minimizes the stripping of potentially on the foaming behavior influential components. For the characterization of the foam stability by this method, Bikerman suggests the calculation of a characteristic number, the so-called "unit of foaminess" (Bikerman coefficient) Σ [20].
This number is determined experimentally and is defined as the ratio of the generated foam volume and the gas flow rate:

$$\Sigma = \frac{V_{foam}}{\dot{V}}$$

Since foams are complex and the description of their behavior is only in a few cases possible, all potential influencing variables such as pressure, temperature, purity of materials etc. should correspond with the respective application.

Experimental examination of fluid dynamics

Due to an ambiguous column design under foaming conditions, one project's objective is the development of a foam test cell adapted to packed columns. This test cell is capable of various process parameters such as hydrodynamics, gas and liquid flow rates, temperature, etc. In the case of a potential foam problem, test cell experiments are realized, which can be performed with the real feed at process conditions of the production column. If a material system is identified in the test cell as problematic, this can be taken into account for an appropriately process design.

To ensure the transferability to columns, an appropriately sized pilot plant for the examination of the fluid dynamics of foaming material systems was constructed at the department. The column with a diameter of 300 mm and 2,000 mm packing height for the observation of foam phenomena is made entirely of glass. As shown in the P&ID diagram (Figure 4), the recycled gas flow in the fully automated pilot plant is controlled with a blower.
Seven commercial packings (random and structured, Table 1) are compared according to their respective sensitivity to foaming with several foaming material systems. The operation experiences should provide hints concerning advantageous packing parameters and are used for the upscale of the measurement results of the adapted foam test cell.

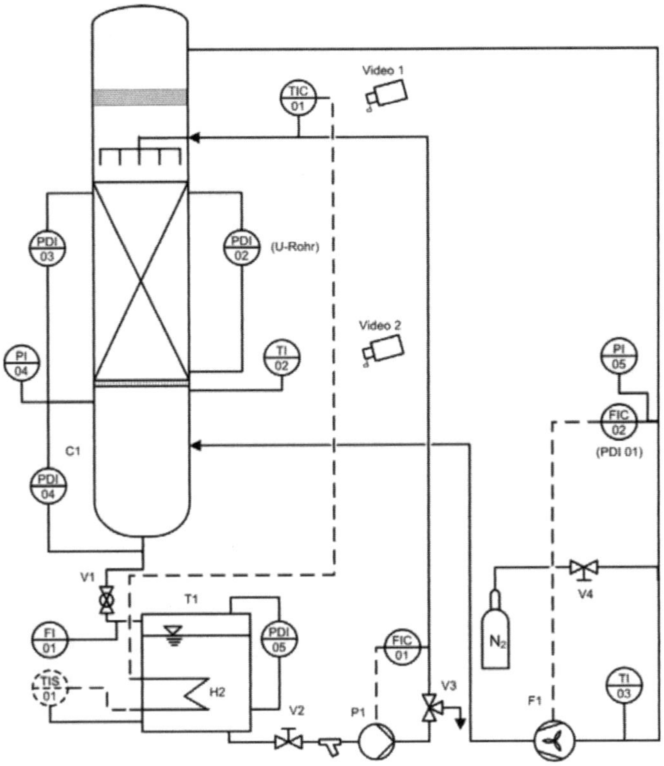

Fig. 4: P & ID - diagram of the pilot plant with 300 mm column

Tab. 1: Examined packings (a = specific surface)

Packing	Type	a [m²/m³]	Material
Montz-Pak B1-350M	structured packing	350	stainless steel
Rauschert Hiflow 15-7	closed random packing	313	polypropylene
Rauschert Hiflow 20-4		280	ceramic
Envimac Mc-Pac #0,75	open random packing	220	stainless steel
Raschig Super-Ring #0,3		315	
Raschig Super-Ring #1		160	
Raschig Super-Ring #2		97,6	

The fluid dynamic of the system butanol-water / air was examined for all packings at liquid loads of L = 0; 10; 30 and 50 m³/m²h. The foaming behavior of the aqueous solution of 2 weight percent butanol has a low foam decay time and a relatively good foaming reproducibility. Furthermore, by monitoring the process, places of foam formation were identified. As a non-foaming reference system, the results are compared with the system water / air.
It has been found, that even at low specific packing pressure losses flooding phenomena appeared far below reference load limits.

Flooding phenomena were observed within the column at several places. In the case of a structured packing (B1-350M), foam is predominantly generated at the segment transitions. This could be explained by the shear stress induced by the forced deflection of both phases.

For very low liquid loads (10 m³/ m²h) there were no operation limitations with the system butanol-water / air for any packaging at all. At low liquid loads of 10 and 30 m³/m²h, generated foam accumulates primarily in the upper packing. Since the liquid distributor reduces the free cross-section, the higher local gas velocity inhibits entrained liquid of returning towards the packing. An increasing foam accumulation in this area can lead to distributor flooding. This was the case in the high performance structured packing B1-350M and random packings Hiflow ring 15-7 and 20-4.

Since the specific packing pressure drop does not change significantly, this parameter doesn't indicate the flooding of the distributor. In industrial operation, the contamination of the head product through massive entrainment could be unnoticed.
With the Super-Rings #0.3 and #1 flooding of the packing is initiated from the supporting grid. Furthermore, a pulsating flooding was observed with the Super-Ring #0.3. The liquid is accumulating in the lower packing, the pressure loss increases and then the pressure is relieved in cyclical waves. The pressure loss curve for this operating condition is shown in Figure 5.

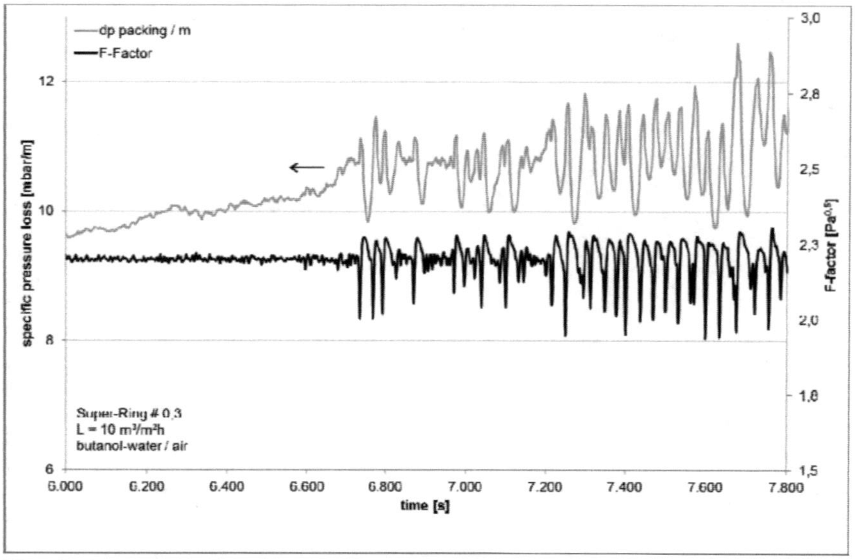

Fig. 5: Periodical flooding (Super-Ring #0.3)

At higher liquid loads (30; 50 m^3/ m^2h) the operation is problematic because of an increase of occurring foam. For the liquid load 30 m^3/ m^2h the operation of the packing B1-350M and Hiflow ring 15-7 is limited, while the Hiflow ring 20-4 is not operable at all. In addition, the reproducibility of the specific pressure loss and flood point for some packings at this liquid load were fading.
At a liquid load of 50 m^3/m^2h, foam accumulates with this experimental set-up mainly in the column bottom and floods the gas inlet (Figure 2). The continuous liquid flow stabilizes the foam in the bottom. In addition, foam is generated by the reflux impact and gas injection at the liquid level. The liquid load seems to have a greater impact on foaming than the gas load.

In respect of the specific pressure loss causing flooding, slightly increasing pressure losses over several hours as well as abrupt pressure losses were observed. Figure 6 shows such an abrupt increase in pressure loss for a structured packing.

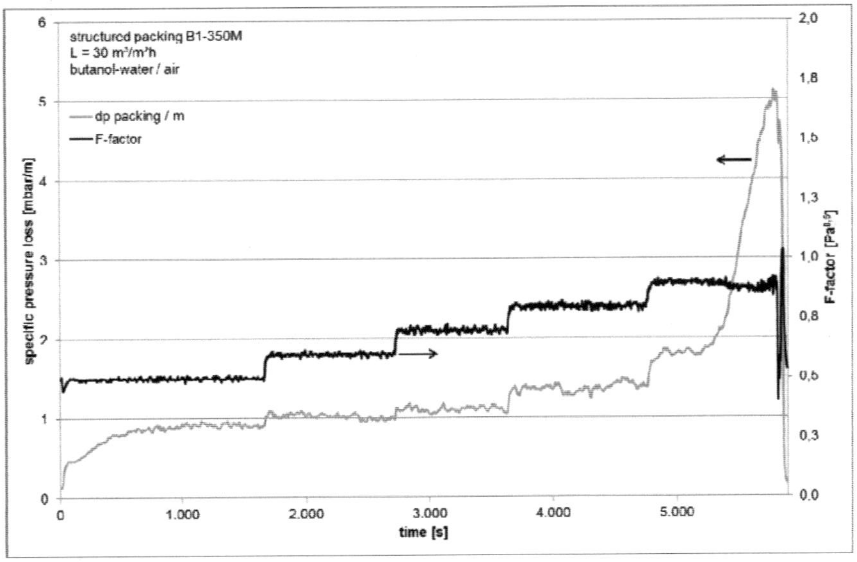

Fig 6: Abrupt pressure loss increase (butanol-water / air, B1-350M, L = 30 m³/m²h)

The specific pressure loss profiles as a function of the F-factor are for low gas loads similar to non-foaming systems. Figure 7 shows an example of a pressure loss characteristic of a random packing for foaming compared to non-foaming conditions.

While the specific pressure loss for low liquid loads as 10 m³/m²h is for foaming compared to a non-foaming system for the examined packings nearly identical, the pressure loss under foaming conditions increases strong for the liquid loads (30 m³/m²h and above). As shown in Figure 7, there is no significantly greater curve increase in the log(Δp)-log(F-factor) diagram, as known from non-foaming systems above the stagnation point. Therefore, a direct comparison of a flood point shift is difficult, especially since foam can flood the liquid distributor quite fast.

For foaming systems, flooding occurs at a much lower specific pressure loss and in most cases abruptly. It has been shown that the pressure loss development do not directly give hints on the load limits and safe operation conditions. An estimation of an acceptable operating range is therefore not achievable by the pressure loss characteristic.

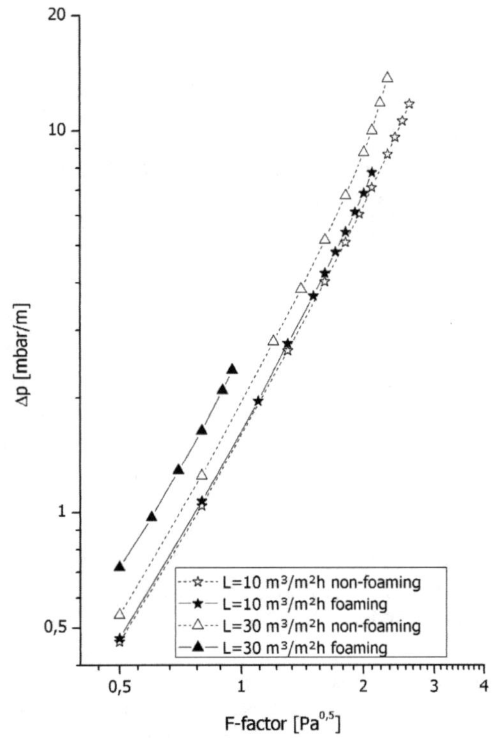

Fig. 7: Pressure loss comparison for Super-Ring #0.3 under foaming (butanol-water / air) and non-foaming (water / air) conditions

Development of an adapted test cell

Various test cells for the indication of the foaming behavior are known. In mass transfer columns with complex geometries (internals), fluid dynamic shear forces effect foam production and destruction. In addition, the liquid reflux stabilizes existing foam. Since known qualification methods do not include these factors, their transferability appears inappropriate in regard to columns. In addition, the test procedures are largely tied to environmental conditions and cannot take changes in phase compositions in account.

An adapted foam test cell is developed, which is capable of foam generation by aeration and by a height-adjustable sprinkler. The set-up shown in Figure 8 is based on a temperature-controlled tank with 80 mm diameter in which a liquid sample can be frothed up by aeration through a glass frit and by sprinkling.
The set-up has a wide operating range of (up to 200 °C, vacuum) and is resistant to many material systems. The investigation of insertable packing geometries allows the consideration of shear force influence on the foaming behavior.

With a foam test cell according to Bikerman as well as with the adapted test cell experimental investigations with the two systems 2 weight percent butanol-water and 40 weight percent MDEA-water were carried out. MDEA is an industrial solvent for CO_2 capture.

Fig. 7: Adapted test cell Ø 80 mm with adjustable heating and liquid sprinkling

In the Bikerman set-up with gassing by nitrogen through a glass frit, both liquids were foaming. While the MDEA solution - similar to butanol-water - was foaming heavy in the Bikerman test cell, an unproblematic pilot plant operation was possible - in contradiction to the butanol solution. A prediction of the foam tendency based on aerating (Bikerman set-up) seems therefore not sufficient enough for a robust applicability to columns.
In the adapted test cell, the foam generation by sprinkling was found more reliable in terms of transferability to packed columns. In that case, only the irrigated butanol-water solution was significantly foaming. In addition, experiments with simultaneous aeration and sprinkling were done. It was observed that foaming of the butanol solution, which foamed heavy in the pilot plant, foamed significantly more than the slightly foaming MDEA solution.
It appears that a prediction based on irrigation enables a better transferability to column operation than on aeration. Therefore, this aspect is currently being investigated.

Conclusions

During the operation of mass transfer columns, foam increases the pressure loss and entrainment and therefore shifts flooding to lower loads. Those means reduce the column capacity significantly.

The aim of the experimental investigations is a safe prediction of foaming behavior and its impact on the fluid dynamics in packed columns. Current foam test methods are not able to predict the foam tendency in relation to packed columns. This will be achieved by systematic investigations on a pilot plant with various packings and test systems as well as the implementation of an appropriate new test cell test to determine the foaming tendency of material systems. It has been shown that a classification of the foam tendency in a test cell is possible. Similar to the conditions in columns, sprinkling a gasified liquid in a test cell provides more durable insights as in the case of foaming due to aeration.

The authors gratefully acknowledge the financial support for the project "foaming media in packed columns" (AiF-no. 16073) granted by the "Applied Industry Research Foundation" (AiF) and by the collaborating enterprises.

Glossary

a	=	specific surface [m²/m³]
p	=	pressure [Pa]
r	=	curvature radius [m]
V_{foam}	=	generated foam volume [m³]
$\dot{V}$	=	gas load [m³/s]
Π	=	disjoining pressure [Pa]
σ	=	surface tension [J/m²]
Σ	=	Bikerman index [s]

References

[1] H.Z. Kister, What Caused Tower Malfunctions in the Last 50 Years?, *Chem. Eng. Res. Des.* **2003**, 81 (1), 5-26, DOI: 10.1205/026387603321158159

[2] R. Thiele, J.-U. Repke, B. Diekjakobs, H. Thielert, G. Wozny, A General Rate-Based Model for Industrial Reactive Absorption and Desorption Processes in Sour Gas Treatment, Präsentation auf AIChE Spring Meeting: Distillation **2005**, Topical Conference Proceedings, 197-208

[3] Y. Zheng, S. Kiil, J.E. Johnsson, Experimental investigation of a pilot-scale jet bubbling reactor for wet flue gas desulphurization, *Chem. Eng. Sc.* **2003**, 58 (20), 4695-4703, DOI: 10.1016/j.ces.2003.07.002

[4] D. Wasan, A. Nikolov, A. Shah, Foaming-Antifoaming in Boiling Suspensions, *Ind. Eng. Chem. Res.* **2004**, 43 (14), 3812-3816, DOI: 10.1021/ie0306776

[5] H. Z. Kister, Distillation Operation, McGraw Hill, New York **1990**

[6] C. Fehn, Untersuchungen zur Stabilität von Tensidschäumen, *Dissertation*, TU Bayreuth **2006**

[7] J. A. F. Plateau, Statique Expérimentale et Théorique des Liquides soumis aux seules Forces Moléculaires, Gauthier-Villars, Paris **1873**

[8] S. Ross, Mechanisms of Foam Stabilization and Antifoaming Action, *Chem. Eng. Prog.* **1967**, 63 (9), 41-47

[9] H. Lange, Schäume und ihre Stabilität, VDI-Berichte **1972**, 182, Gemeinsame Tagung über Blasen und Schäume, VDI-Verlag, Düsseldorf, 71-77

[10] S. Ross, G. Nishioka, Foaminess of Binary and Ternary Solutions, *J. Phys. Chem.* **1975**, 79 (15), 1561-1565, DOI: 10.1021/j100582a018

[11] M.J. Lockett, Distillation Tray Fundamentals, University Press, Cambridge **1986**

[12] M. H. Pahl, D. Franke, Schaum und Schaumzerstörung – ein Überblick, *Chem. Ing. Tech.* **1995**, 67 (3), 300-312, DOI: 10.1002/cite.330670306

[13] N. D. Denkov, Mechanisms of Foam Destruction by Oil-Based Antifoams, *Langmuir* **2004**, 20 (22), 9463-9505, DOI: 10.1021/la049676o

[14] W.L. Bolles, The Solution of a Foam Problem, *Chem. Eng. Prog.* **1967**, 63 (9), 48-52

[15] C. Marangoni, Über die Ausbreitung der Tropfen einer Flüssigkeit auf der Oberfläche einer anderen, *Annalen der Physik und Chemie* **1871**, 143 (7), 337-354

[16] N. Lebedev, A. I. Vladimirov, V. D. Kos'min, Effect of Foaming on Hydrodynamics of Mass-Transfer Contact Devices, *Chem. Technol. Fuels Oils* **1997**, 33 (6), 332-334

[17] R. Thiele, O. Brettschneider, J.-U. Repke, H. Thielert, G. Wozny, Experimental Investigations of Foaming in a Packed Tower for Sour Water Stripping, *Ind. Eng. Chem. Res.* **2003**, 42 (7), 1426-1432, DOI: 10.1021/ie020676y

[18] R. Thiele, H. Wiehler, J.-U. Repke, H. Thielert, G. Wozny, Hydrodynamics of Foaming Systems in Packed Towers, Präsentation auf AIChE Annual Meeting **2004**, Austin

[19] G. X. Chen, T. J. Cai, K. T. Chuang, A. Afacan, Foaming Effect On Random Packing Performance, *Chem. Eng. Res. Des.* **2007**, 85 (2), 278 - 282, DOI: 10.1205/cherd06095

[20] J.J. Bikerman, Foams, Springer, New York **1973**

[21] R. K. Prud'homme, S. A. Khan, Foams – Theory, Measurements, and Applications, *Surfactant Science Series*, 57, Marcel Dekker, New York **1996**

(102d) Distillation Column Flooding Predictorô - Increase Throughput, Improve Energy Efficiency, and Avoid Flooding

George Dzyacky and Steven Carlson, 2ndpoint, LLC, Schererville, IN

Distillation Column Flooding Predictor™ - Increase throughput, improve energy efficiency, and avoid flooding

The Flooding Predictor™ is a patented advanced control technology proven in research at the Separations Research Program, University of Texas at Austin, to increase distillation column throughput by over 6%, while also increasing energy efficiency by 10%.

The research was conducted under a U. S. Department of Energy Cooperative Agreement awarded to George Dzyacky of 2ndpoint, LLC. The Flooding Predictor™ works by detecting the incipient flood point and controlling the column closer to its actual hydraulic limit than historical practices have allowed. Further, the technology uses existing column instrumentation, meaning no additional refining infrastructure is required.

Refiners often push distillation columns to maximize throughput, improve separation, or simply to achieve day-to-day optimization. Attempting to achieve such operating objectives is a tricky undertaking that can result in flooding. Operators and advanced control strategies alike rely on the conventional use of delta-pressure instrumentation to approximate the column's approach to flood. But column delta-pressure is more an inference of the column's approach to flood than it is an actual measurement of it. As a consequence, delta pressure limits are established conservatively in order to operate in a regime where the column is never expected to flood. As a result, there is much "left on the table" when operating in such a regime, i.e. the capacity difference between controlling the column to an upper delta-pressure limit and controlling it to the actual hydraulic limit. The Flooding Predictor™, an innovative pattern recognition technology, controls columns at their actual hydraulic limit, which research shows leads to a throughput increase of over 6%.

Controlling closer to the hydraulic limit also permits operation in a sweet spot of increased energy-efficiency. In this region of increased column loading, the Flooding Predictor is able to exploit the benefits of higher liquid/vapor traffic that produce increased contact area and lead to substantial increases in separation efficiency – which translates to a 10% increase in energy efficiency on a BTU/bbl basis.

The Flooding Predictor™ operates on the principle that between five to sixty minutes in advance of a flooding event, certain column variables experience an oscillation, a pre-flood pattern. The pattern recognition system of the Flooding Predictor™ utilizes the mathematical first derivative of certain column variables to identify the column's pre-flood pattern(s). This pattern is a very brief, highly repeatable, simultaneous movement among the derivative values of certain column variables. While all column variables experience negligible random noise generated from the natural frequency of the process, subtle pre-flood patterns are revealed among sub-sets of the derivative values of column variables as the column approaches its hydraulic limit.

The sub-set of column variables that comprise the pre-flood pattern is identified empirically through in a two-step process. First, 2ndpoint's proprietary off-line analysis tool is used to mine historical data for pre-flood patterns. Second, the column is flood-tested to fine-tune the pattern recognition for commissioning. Then the Flooding Predictor™ is implemented as closed-loop advanced control strategy on the plant's distributed control system (DCS), thus automating control of the column at its hydraulic limit.

See more of this Session: Contributions to Distillation Troubleshooting and Control
See more of this Group/Topical: Topical 8: The Dr. James Fair Heritage Distillation Symposium

Improved Debutanizer Quality Control using Weighted Average Temperature

Karl-Martin Hansson, Nicholas Alsop, Linda Fredriksson
PREEM AB, Preemraff Lysekil SE-453 81, SWEDEN

Abstract

This article describes a successful implementation of a control scheme based on weighted average temperature for improving the stability of product qualities in a debutanizer at Preem Refinery in Lysekil Sweden.

The original control scheme for the 30 tray debutanizer tower was based on a temperature controller using the tray # 8 temperature to control the split between C_{4-} and naphtha. The control never worked and for many years the top LPG-draw flow rate was controlled manually by the control room operators. Poor control of the cut-point resulted in, on average, fairly high concentrations of pentane in the LPG and of butane in the naphtha, even though separation in the tower was good at all times.

The reason the original control scheme failed to control the cut-point was due to the highly non-linear relationship between tray # 8 temperature and the flow used to control the temperature. Tray # 8 temperature is highly sensitive to the LPG-draw when there are pentanes in the top, but insensitive to the LPG-draw when the draw is too low at which times significant amounts of butane end up in the bottom of the tower.

It was noted by control room operators that the temperature on tray # 24 is very sensitive to the concentration of butane in the bottom. Even so, the temperature on tray # 24 is not suited for cut-point control due to poor sensitivity when the LPG-draw is too large. Therefore neither temperature is suitable for cut-point control.

The solution to the control problem was to use a calculated control temperature that weighted the temperatures on tray # 8 and tray # 24. The temperature on tray # 8 indicates when the LPG-draw is too large whereas the temperature on tray # 24 picks up if the LPG-draw is too small. Using the calculated control temperature, it is possible to attain on average lower concentrations of pentane in the LPG and of butane in the naphtha, using a lower reboiler duty than before.

1 Background

1.1 Process description

The debutanizer is fed with naphtha from a desulphurisation unit. There are two top products; off-gas sent either to an LPG recovery unit or to the fuel gas system, and a liquid LPG-stream (butane and lighter) sent to an LPG separation unit. The bottom product is stabilised naphtha sent in turn to a naphtha splitter. There are 15 trays in the rectifying section and 15 trays in the stripping section. Reboiler duty is supplied in the convection part of one of the reformer heaters (duty not controlled) and then in a dedicated heater. The feed rate of naphtha to the unit is normally 280-350 Sm^3/h. The debutanizer normally operates at a reflux rate of about

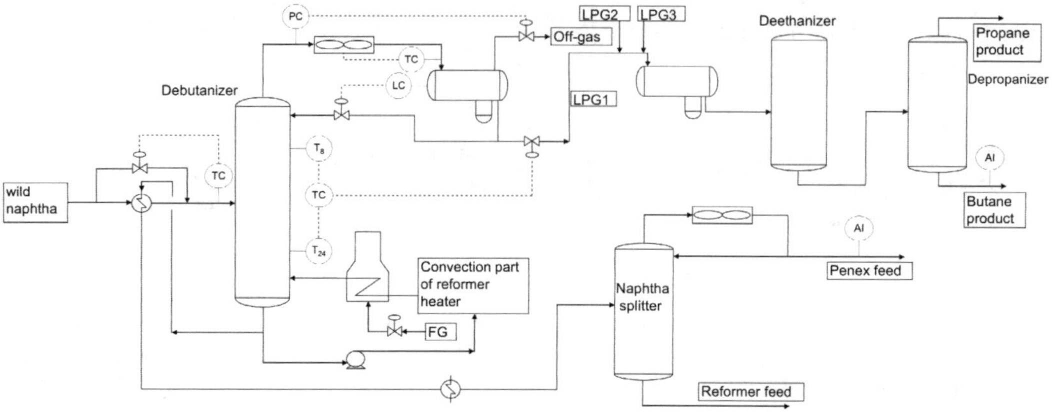

Figure 1. The debutanizer with its new temperature controller, naphtha splitter and LPG separation unit (deethanizer + depropanizer).

150 Sm^3/h and the LPG draw rate is typically 3-12 Sm^3/h. The measured off-gas flow rate is approximately 3000 Sm^3/h. Temperatures are measured at both trays # 8 and # 24 (counting from the top) as well as in the top and in the bottom product line.

The primary quality specification is the composition of C_5 in the top liquid product, which is based on the specification for C_5 in butane product. The secondary quality specification is the composition of C_4 in the bottom product. There is an analyser for C_4 in the top of the downstream naphtha splitter (see Figure 1). From this analyser the quantity of C_4 in the naphtha from the debutanizer can be deduced, but due to the long dead time (one tower downstream and slow analysis) associated with the analyser, it is unsuitable for online control. There is also an analyser for C_5 in the bottom stream from the depropanizer in the LPG separation unit (one surge drum and two towers downstream, see Figure 1). There are two reasons why this analyser is not suitable for online control. The first reason is the long dead time between the debutanizer and the analyser. The second reason is that the LPG unit is fed with LPG from 2 other sources besides the debutanizer in question and a peak of C_5 in the butane product can originate from any of the other 2 sources. For one of the other streams, there is another online analyser, but the other stream lacks an analyser.

The average concentration of C_5 in the LPG (from routine sampling) used to be 1 mole-% which corresponds to 1.5 mole-% of the C_{4+} fraction of the LPG. The highest recorded C_5 concentration in the LPG from routine sampling is above 12 mole-% (16% of the C_{4+}). The product specification for C_5 in butane product is 2 mole-%. Though the average concentration of C_5 in the C_{4+} fraction of the LPG of the debutanizer top product and other butane streams in the refinery are within the product specification, peaks in the C_5 impurity level must be avoided in order to keep the blended butane within product specification at all times.

There is no product specification for butane in the naphtha, but the top product from the naphtha splitter is sent to a Penex unit where the butanes take up valuable space and from where the butanes are finally driven off and sent to the fuel gas system. Excess butanes in the bottom from the debutanizer are thus wasted as fuel gas. The average concentration of butanes in the top product from the naphtha splitter (some 30% of the total naphtha flow) used to be 1.1 vol% (corresponding to more than 1 Sm^3 of butanes/h on average).

1.2 Description of the original control system

In the original control system, top liquid product flow was used to control level in the overhead accumulator and reflux was used for control of temperature on tray # 8 (Figure 2.A). This control scheme proved highly unsatisfactory and resulted in frequent loss of level in the accumulator. The problem was due to the small top liquid product flow in relation to the total liquid throughput in the accumulator drum. A simple remedy to this problem was to control accumulator level with reflux and temperature using LPG draw rate (Figure 2.B).

Despite the vast improvement in accumulator level stability, the temperature controller in Figure 2.B was unable to maintain set-point despite countless tuning efforts. The operators had no confidence in the controller and for many years the top LPG draw rate was instead put on manual control. The poor control of the cut-point thus achieved resulted in on average fairly high concentrations of pentane in the LPG and of butane in the naphtha, even though separation in the tower was good at all times.

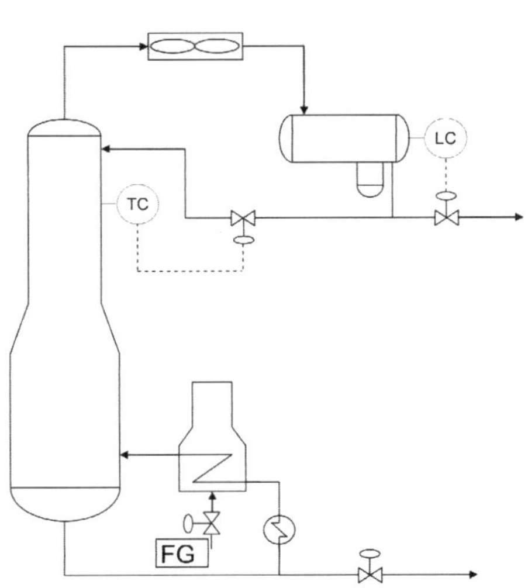

Figure 2.A. First control configuration – frequent loss of accumulator level.

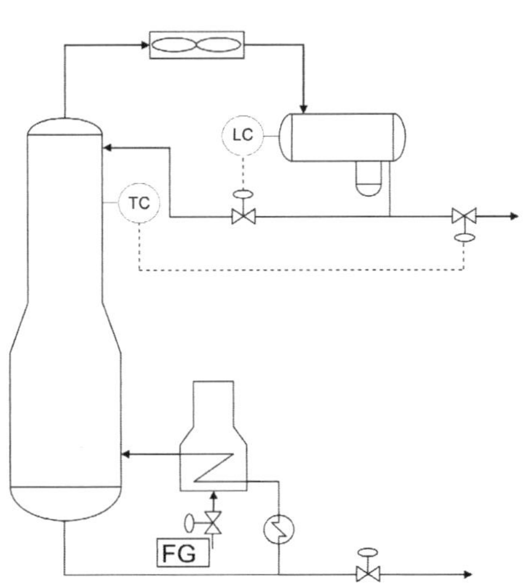

Figure 2.B. Second control configuration – cut-point control did not work.

1.3 Control room operator experience

Some of the control room operators soon learned to see whether the top liquid draw rate was too small, too large or just about right. These operators noted that before the C_4 concentration in the naphtha splitter overhead started to increase, the temperature on tray # 24 in the debutanizer would dramatically decrease. Also, they noted that as the temperature on tray # 8 in the debutanizer started to increase rapidly, C_5 concentration in the butane product would soon follow. These control room operators used the temperatures in the column as their primary indicators. However, other operators used the online analysers as their only means to control the LPG draw rate. While the C_4 analyser for the naphtha splitter overhead stream was suitable for control guidance, the C_5 analyser for product butane was not. As mentioned, the LPG separation unit was also fed with 2 other LPG-streams, one stream from the crude naphtha stabiliser and one stream from a reformate debutanizer. On some occasions when the C_5 concentration in the product butane was high, this was due to a high concentration of C_5 in the LPG from the crude unit (as seen on its online analyser) or due to high concentrations of C_5 in the LPG from the reformate debutanizer (as seen from high top temperature in the reformer debutanizer). Even so, operator action was often taken on the prime suspect, the debutanizer in question.

Even for the control room operators who knew what to look for, it was difficult to control the LPG draw rate. First of all, they had a lot of units to monitor and secondly, the response time for the debutanizer is several hours. If a corrective measure had been taken and the response had not followed after an hour, an additional corrective measure often followed. Since the effect of the first action had not yet been fully attained, the quality targets were frequently overshot.

1.4 Analysis of historical data

An attempt to understand the tower behaviour was made in 2006. First of all, operations data were checked and the observations of some of the control room operators were confirmed. Figure 3 shows how the temperature on tray # 24 in the tower varied with time during a typical week in 2006. Also shown in Figure 3 is how the concentration of n-butane in the naphtha splitter overhead varied during that week. It is apparent that a drop in the temperature on tray # 24 was followed by a lot of n-butane (over 3 volume-% on day 4) in the naphtha splitter overhead liquid. As long as the temperature on tray # 24 was between 165-170 °C there was no butane in the bottom of the debutanizer.

Temperatures on tray # 8 in the debutanizer in the same time period are shown on the right in Figure 3. Every time the temperature on tray # 8 was higher than 90-100 °C, the concentration of i-pentane in the butane product from the depropanizer in the LPG separation unit peaked. There are also peaks in i-pentane that are associated with too high top temperatures in the reformate debutanizer. The concentration of i-pentane in the LPG from the crude stabilizer unit ranged between 0.35 and 1.15 % according to its online analyser. Since the crude LPG contained propane as well as C_{4+}, the C_5 concentration in the C_{4+} fraction was higher than the analysed C_5 concentrations in the whole LPG-stream from the crude unit.

Most of the incidents when the temperatures on trays # 8 or # 24 started to deviate from their ideal values were associated with crude switches. Every time the crude is switched the naphtha rate and naphtha composition from the crude unit changed and the amount of C_4 in the naphtha from the crude naphtha stabiliser changed. Some disturbances in the debutanizer in question also arose from operational changes in the crude naphtha stabiliser (changes in cut-point) or from changes in naphtha feed rate.

Figure 4 summarises the operating regimes observed in the historical data in the period chosen. As shown, the temperature on tray # 8 did not drop below 85 °C, no matter how far the temperature on tray # 24 dropped as more butane was let out in the bottom of the debutanizer. This explains why the original control scheme did not work in that the temperature on tray # 8 is very sensitive to the LPG-draw as long as the draw rate is too high, but completely insensitive to the draw rate when the draw rate is too low. The temperature on tray # 24 never went above 170 °C regardless if the temperature on tray # 8 was as high as 115 °C, corresponding to excess LPG draw rate and C_5 in the LPG. Thus, the temperature on tray # 24 is equally unsuited for proper control of the cut-point.

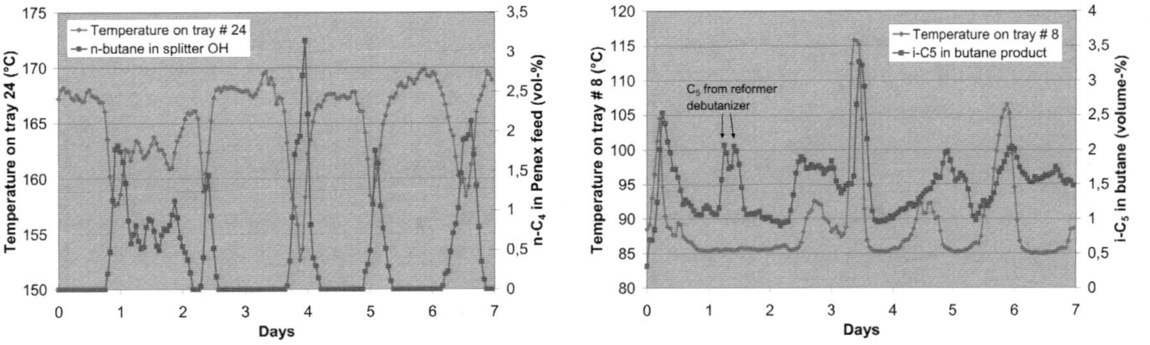

Figure 3. Left: The temperature on tray # 24 in the debutanizer and concentration of n-butane in the naphtha splitter overhead. Right: The temperature on tray # 8 in the debutanizer and concentration of i-pentane in the butane product from the LPG depropanizer. Average values for the week were; reflux=155 Sm^3/h, feed to the naphtha unit reactors=301 Sm^3/h, LPG draw=9 Sm^3/h, stabilised naphtha flow=306 Sm^3/h. The LPG draw rate varied between 4 and 12 Sm^3/h. Top pressure was 10.7 barg.

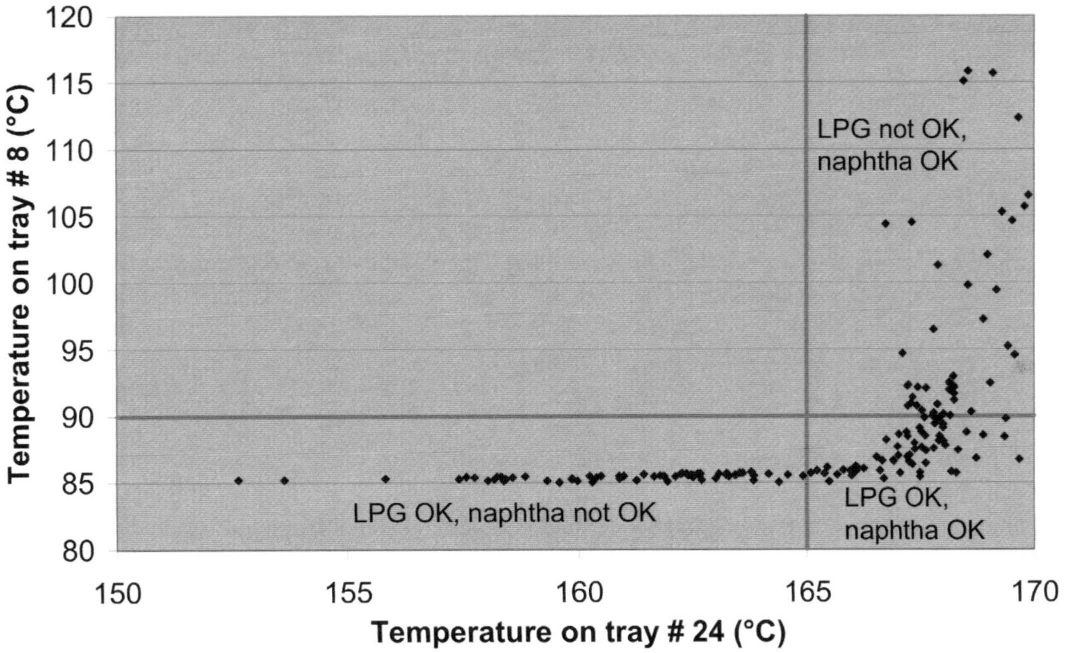

Figure 4. One-hour-averages for temperatures on trays # 8 and # 24 for one week. Also shown are approximate limits for when C_5 in the LPG and C_4 in the bottoms naphtha are on specification. Top pressure was 10.7 barg.

2 Revised Temperature Control Scheme

As neither tray # 8 nor tray # 24 temperatures were suitable for temperature control, a solution to the temperature control problem was proposed on the basis of a calculated temperature. The calculated temperature is the weighted average of the measured temperatures on tray # 8 and tray # 24 according to:

$$T_c = \eta \cdot T_8 + (1-\eta) \cdot T_{24} \quad (1)$$

Where:
T_c = Calculated temperature to be controlled by LPG draw rate (°C)
T_8 = Temperature on tray # 8 (°C)
T_{24} = Temperature on tray # 24 (°C)
η = Coefficient between 0 and 1

It was recognised that the value of weighing coefficient "η" could be selected using a design procedure similar to that used for selection of control tray temperature [1]. The design procedure is based on finding the temperature which maximises sensitivity, linearity and correlation to product quality. For the debutanizer tower in question, the main problem in selecting control temperature is linearity. Therefore η was selected so as to maximise the linearity in the relationship between the LPG draw rate and the calculated temperature.

Figure 9 shows simulated weighted temperatures as a function of LPG draw rate for three different weighing coefficients ranging from 0.5 to 0.1. As can be seen in the figure, the most linear response is obtained by using a weighting coefficient of 0.1. This visual result was verified using a spreadsheet procedure which maximised the statistical coefficient of correlation (R2) between LPG draw rate and temperature for varying levels of η. The result shows that the temperature response is most linear and therefore easiest to control when the weighed average favours strongly the temperature on tray # 24. The risk with this weighing is that a disturbance which lowers the temperature on tray # 24 by 1 °C must be compensated by an increase of 9 °C on tray # 8 in order for the control temperature to be kept at the same value.

The effect on different conceivable disturbances in feed composition and feed rate were estimated in simulations (see Section 3). It was found that changes in feed composition would not affect the ideal weighted temperature by much, but that changes in reboiler duty/feed rate would have some effect. It was considered that since the only disturbances that would require an operator response (change in control temperature set-point) were when the naphtha feed rate or the reboiler duty was changed significantly. As both of these follow operator action, it is reasonable that the weighing favours linearity, especially since the most common disturbance is butane concentration in the feed. According to the simulation, the weighted control temperature is insensitive to this disturbance. The fact that there are online instruments that would indicate if the control temperature set-point is badly chosen justified our design decision on the basis of linearity.

The temperature controller using the calculated T_c with η=0.1 was built in the Honeywell TDC. Aggressive use was made of derivative action in the PID controller to accommodate for the high order process dynamic in the response between LPG draw rate and calculated temperature. The revised temperature control scheme based on calculated temperature was taken in service in 2009 and worked very well as demonstrated by a reduction both in C_4 composition in the bottom product and C_5 in the LPG.

3 Simulations

Ideally a test run was required to see the temperature response on variations in LPG draw rate. However, the frequent changes in LPG in the feed to the unit in combination with very long times to reach steady state made this practically impossible in 2006. Instead a simulation study was performed to estimate how the tower temperatures varied with LPG draw rate. This study was used in choosing the weighing coefficient η. The impact of various disturbances including variations in the feed composition to the unit were also estimated.

3.1 Simulation input – base case simulation

AspenTech's simulation package "HYSYS" was used. The input data to the simulation comprised laboratory analyses of off-gas, LPG and bottoms product taken on an arbitrarily chosen day. The refinery laboratory analysed all samples using gas-chromatography. The H_2S content in the off-gas was measured separately using a dräger tube.

For the bottom's sample, the laboratory felt confident that all components up to C_8 were accurately identified and quantified. For the heavier part of the naphtha, the resolution of the GC-spectrum was too low and due to overlapping peaks it was not possible to make an accurate quantification.

The C_{5+} in the naphtha from the analysis was 295.5 Sm^3/h. 53.4 volume% of these were made up of identified C_5-C_7 components and these were specified in the simulation. The C_{8+} was lumped together into four n-parafins, arbitrarily assumed to consist of 30 vol% n-octane, 30 vol% n-nonane, 30 vol% n-decane and 10 vol% n-undecane. The assumption that the heavy components in the naphtha could be lumped in this way was checked in the sensitivity analysis (see below).

The total input of C_{5-} to the debutanizer from the sampling is summarized in Table 1.

Soave-Redlich-Kwong's equation of state was chosen for the simulations.

The number of theoretical trays was adjusted in order to achieve a degree of separation that matched actual separation. The simulation package uses a theoretical step for the reboiler, whereas the actual reboiler uses forced circulation to charge a liquid flow more than 2 times the bottom product flow rate, from the bottom sump through the reboilers. Thus, the real reboiler does not provide a theoretical stage but only a fractional stage. The simulation used 14 theoretical trays to represent the rectifying section and 13 theoretical trays plus one theoretical stage for the stripping section and the reboiler.

Table 1. Input data to simulation. The feed to the tower was the mix of the three analysed product flows.

Component	Component flow rate in off-gas (kg/h)	Component flow rate in LPG (kg/h)	Component flow rate in naphtha (kg/h)
H_2	84	Not analysed	0
H_2S	81	18	0
CH_4	72	2	0
C_2H_6	347	75	0
C_3H_8	734	435	0
i-C_4H_{10}	684	754	13
n-C_4H_{10}	1843	2420	652
neo-pentane	0	2.4	18
i-pentane	0	2.9	9458
n-pentane	0	0.5	12363
Cyclopentane	0	0	1940

Table 2. The simulations base case.

Top pressure	10.0 barg
Pressure in overhead accumulator	9.8 barg
Bottom pressure	10.2 barg
Temperature in overhead accumulator	36 °C
Reflux rate	150 Sm^3/h
Typical naphtha contribution to tower feed	296.6 Sm^3/h of which 295.5 Sm^3/h were C_{5+}
Typical LPG contribution to tower feed	3710 kg/h (composition as in Table 1)
Typical off-gas contribution to tower feed	3845 kg/h (composition as in Table 1)
Specified i-pentane in LPG	0.2 mole-%

The real tray # 8 corresponds to between theoretical trays # 7 and # 8, while real tray # 24 corresponds with theoretical tray # 23.

Basic data for the base case simulation is listed in Table 1 (feed composition) and Table 2 (operational data).

3.2 Simulation results – effect of LPG draw rate on tower temperatures

The calculated reboiler duty for the base case was 16.6 MW and this duty was used in the latter simulation cases where the reflux rate was allowed to vary. Figure 5 shows calculated temperature profiles for the base case as well as for two cases where the LPG draw rate had been increased by 1 Sm^3/h and reduced by 1 Sm^3/h respectively.

An interesting observation for the case where the LPG draw rate was increased by 1 Sm^3/h compared to the base case, was that the bottom flow rate only decreased by 0.65 Sm^3/h. The rest of the LPG increase came from the off-gas which decreased by 184 kg/h.

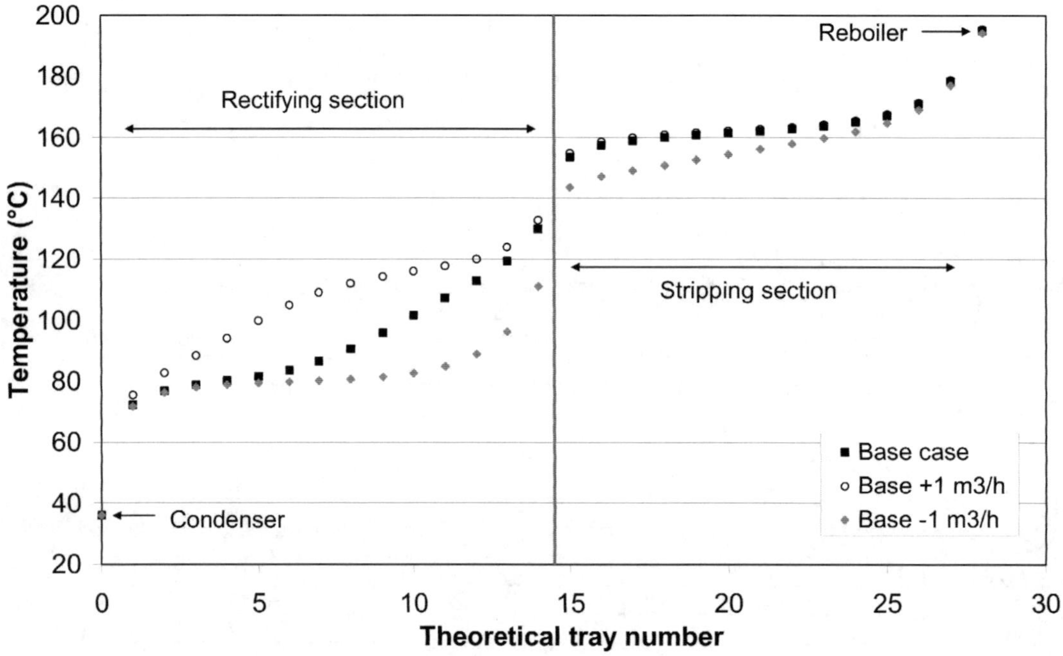

Figure 5. Calculated tower temperatures for the base case and for two cases with the same reboiler duty as for the base case but with LPG draw rates that were plus or minus 1 Sm^3/h relative the LPG draw rate for the base case.

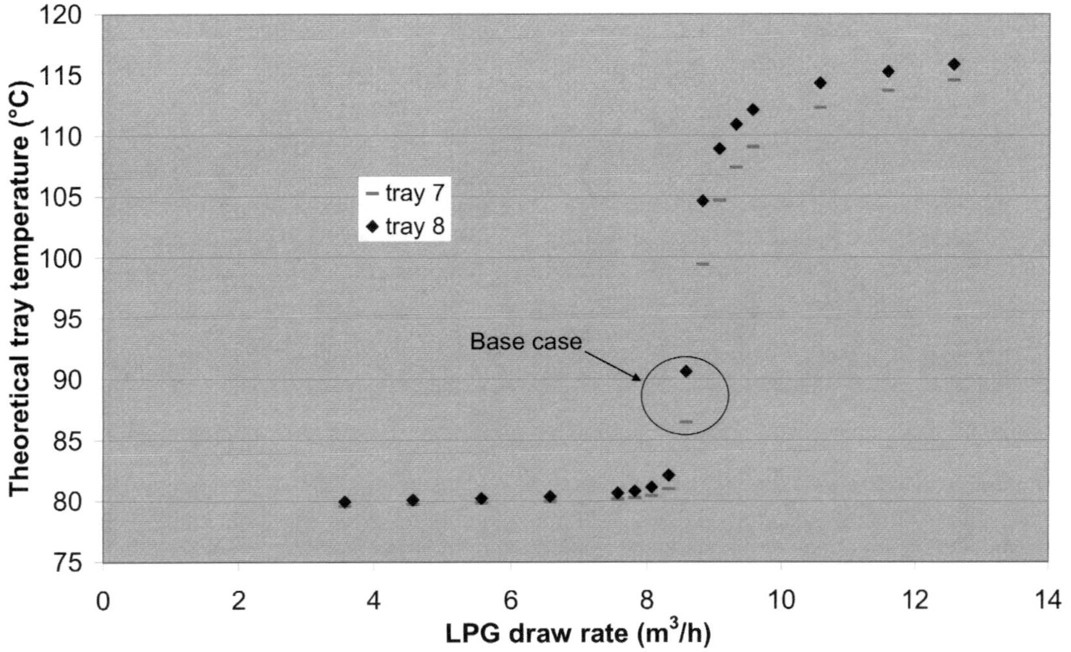

Figure 6. Calculated tower temperatures on theoretical trays # 7 and # 8 as a function of LPG draw rate. Base case and several cases with the same reboiler duty as for the base case but with different LPG draw rates.

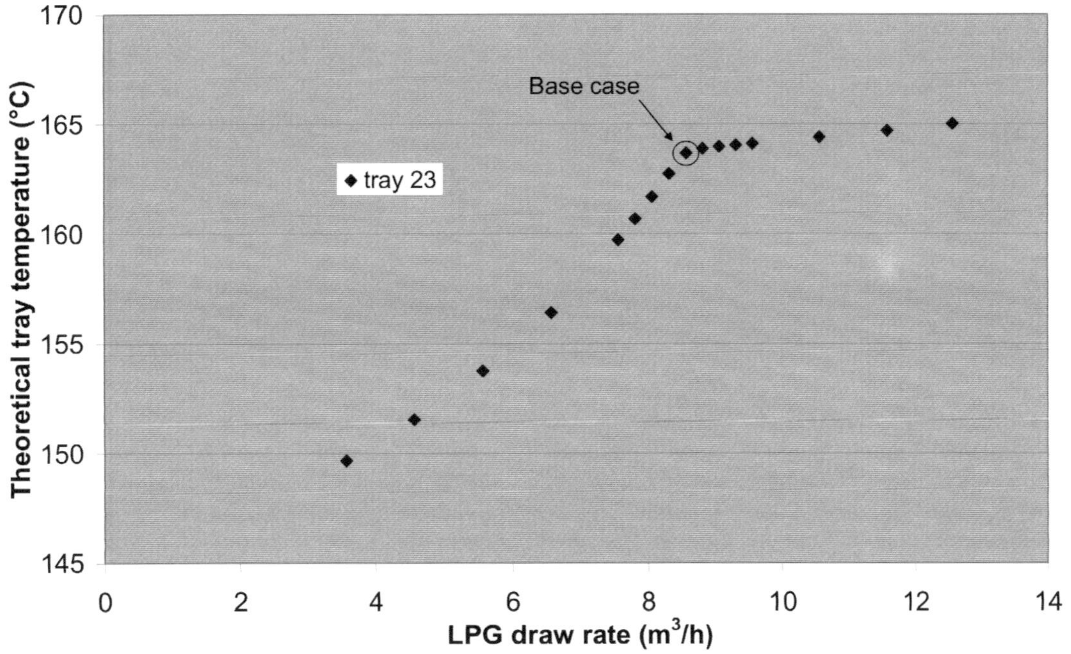

Figure 7. Calculated tower temperatures on theoretical tray # 23 as a function of LPG draw rate. Base case and several cases with the same reboiler duty as for the base case but with different LPG draw rates.

Figure 6 shows the calculated temperatures on theoretical trays # 7 and # 8 (corresponding to real tray # 8) as function of LPG draw rate. Figure 7 shows the effect of LPG draw rate on the calculated temperature on theoretical tray # 23 (corresponding to real tray # 24).

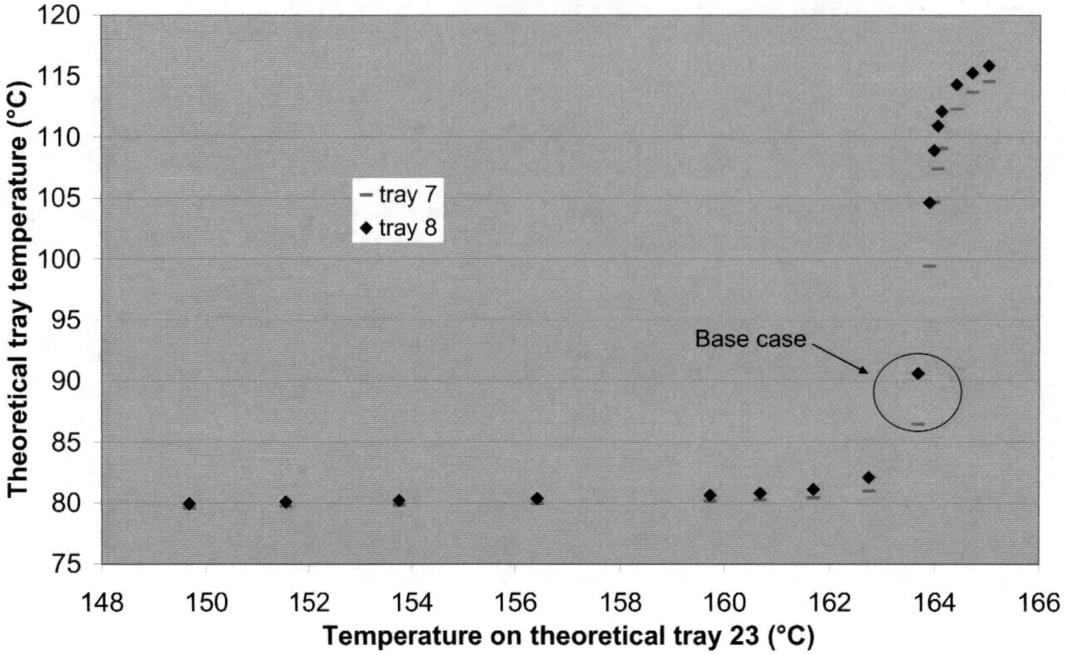

Figure 8. Relationship between theoretical tray #23 and theoretical trays # 7/8 as LPG draw rate varies.

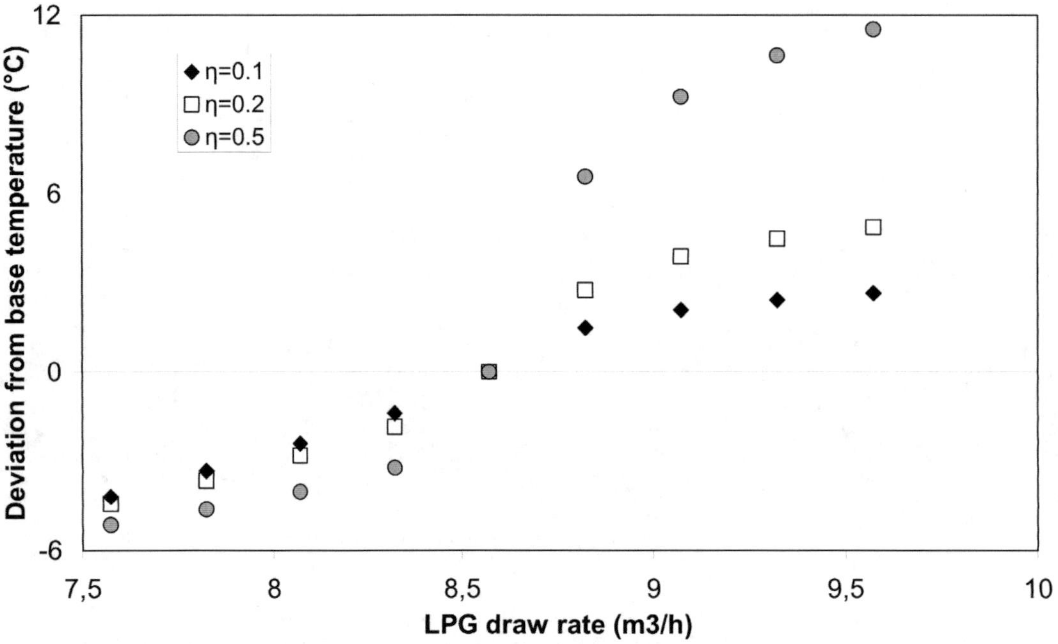

Figure 9. Perturbation in weighted temperature as a function of the LPG draw rate for three different weighing coefficients.

Figure 8 shows how the temperatures on theoretical trays # 7 and # 8 varies with the temperature on theoretical tray # 23 when the LPG draw rate is varied at constant reboiler duty.

Figure 9 shows how different weighted temperatures (weighting using theoretical trays # 7 and # 23) varies with LPG draw rate for three different weighing coefficients. An arithmetic mean average (η=0.5) gives a highly non-linear correlation at the nominal LPG draw rate. The relationship becomes more linear at smaller values of the weighing coefficient.

3.3 Simulation results – sensitivity analysis, increase in lights in tower feed

The butane in the feed to the tower originates from butane in the feed to the HDS reactors, from butane in the hydrogen treat gas used in the HDS and possibly from the desulphurisation reactions. Propane, ethane, methane and hydrogen originates from the hydrogen treat gas, and to some degree from desulphurisation reactions. H_2S originates from the desulphurisation reactions. The flow rates of components lighter than butane depends mainly on the quality of the treat gas used (normally 85-90 mole% hydrogen and 10-15 mole% hydrocarbons) and on the operation of the separators upstream of the debutanizer. The composition is not believed to vary greatly. However, the amount of butane in the naphtha from the crude tower can vary greatly.

Typically, the naphtha feed to the naphtha HDS contains about 2 vol% n-butane and 0.2 vol% i-butane. The effect of adding 3 Sm^3/h n-butane and 0.3 Sm^3/h i-butane in the feed to the unit (corresponding to about 1 vol% more n-butane and about 0.1 more i-butane in the naphtha) was studied using the simulation. Using the reboiler duty and the same weighted temperature (η=0.1) as in the base case, the concentration of i-pentane in the LPG increased from 0.2 to 0.29 mol% while the n-butane in the naphtha increased slightly but remained below 0.05 vol%.

The simulations indicate that the debutanizer can handle more butane in the feed than what is normal with the current reboiler duty with negligible penalty in separation. This means that the crude naphtha stabilizer can be unloaded allowing more butane to remain in the naphtha to the naphtha HDS, which has some over-capacity.

The simulation also suggest that the column was operating at a higher than required fractionation. By specifying 0.3 mole% i-pentane in the LPG and 0.3 vol% of n-butane in the naphtha, would for the base case with reflux and reboiler duty allowed to vary, require a reboiler duty of 12.2 MW corresponding to a reflux of 84 Sm^3/h. At this separation, the LPG draw rate would be almost 1 Sm^3/h less than the base case and the loss in LPG would be counterbalanced by an increase in off-gas.

Though highly unlikely, large variations in components lighter than butane in the tower feed were simulated. Adding 500 kg methane/h to the tower feed and keeping tray # 7 temperature constant leaves tray # 23 temperature practically unchanged. Hence, the same weighted temperature could be used. The off-gas rate increases significantly and the LPG rate decreases by 2.45 Sm^3/h. Reflux remains unchanged and i-pentane in LPG increase from 0.2 to 0.239 mole%. The concentration of n-butane in the naphtha decreased from 0.0350 vol% to 0.0335 vol%. Therefore, an increase in methane has little effect on cut-point. Similarly, increasing the combined feed to the tower by 500 kg/h of ethane or propane had similar – yet less – impact on the cut-point than the increase in methane in that both n-butane in bottom naphtha and i-pentane in LPG remained fairly constant. In summary, variations in lights in the tower feed do not greatly affect the temperatures used in cut-point control.

3.4 Simulation results – sensitivity analysis, naphtha end cut-point

The naphtha cut (D 86, 95%) in the crude tower has decreased by 10 °C over the last 5 years. Hence, one can expect that the naphtha cut may change again. A sensitivity analysis of the impact on the tower temperatures from changes in the heavy end of the feed was performed.

The impact of naphtha cut was checked by rearranging the C_{8+} fraction of the tower feed. Originally the 137.7 Sm^3/h of C_{8+} in the base case was assumed to consist of 30 vol% n-C_8, 30 vol% n-C_9, 30 vol% n-C_{10} and 10 vol% n-C_{11}. By changing this assumed composition to 50 vol% n-C_8, 35 vol% n-C_9, 15 vol% n-C_{10} and 0 vol% n-C_{11} the effect of a lighter naphtha-cut could be simulated. Similarly a heavier naphtha-cut could be simulated by changing the C_{8+} fraction to consist of 10 vol% n-C_8, 30 vol% n-C_9, 30 vol% n-C_{10} and 30 vol% n-C_{11}.

For the heavy naphtha cut, the theoretical tray # 23 temperature dropped by 0.3 °C and the bottoms temperature increased by 1.5 °C when the tray # 7 temperature was kept constant. Concentration of n-butane in naphtha did not change while i-pentane in LPG increased by 1.8% (relative increase). The small increase in i-pentane was due to the 4 Sm^3/h decrease in reflux which followed as more of the reboiler duty was spent on increasing the bottoms temperature. Keeping the weighted temperature constant meant increasing tray # 7 temperature by 2.17 °C and decreasing tray # 23 temperature by 0.24 °C. This resulted in an increase in the concentration of i-pentane in the LPG from 0.2 to 0.3 mole% and a decrease in the n-butane concentration in the naphtha from 0.035 to 0.032 vol%.

The small effects were seen in reverse for the light naphtha cut case. Adjusting tray # 7 temperature so that weighted control temperature was kept constant showed that the naphtha cut has a small effect on the debutanizer cut-point. This means that lumping of heavy components is an acceptable approximation and that the weighed control temperature will not be greatly affected by changes in naphtha cut-point. Furthermore, naphtha cut-point is changed very rarely.

3.5 Simulation results – the effect of reflux ratio

The effect of reboiler duty on product qualities was studied in the simulations. If the concentration of i-pentane in the LPG was to be kept at 0.2 mole%, the theoretical control-temperature (10% of theoretical tray # 7 + 90% of theoretical tray # 23) had to be increased by 3 °C if the reboiler duty was decreased so that the reflux decreased from 150 to 90 Sm^3/h (Figure 10).

As reflux is decreased, the concentration of butane in the bottom naphtha increases as long as the concentration of i-pentane in the LPG is kept constant. For a reflux between 80 and 90 Sm^3/h, the concentration of butane in the naphtha becomes as high as the average value before the controller was taken in service.

The theoretical effect of reflux was found to have such large impact on the choice of control temperature set-point, that this needed to be verified experimentally.

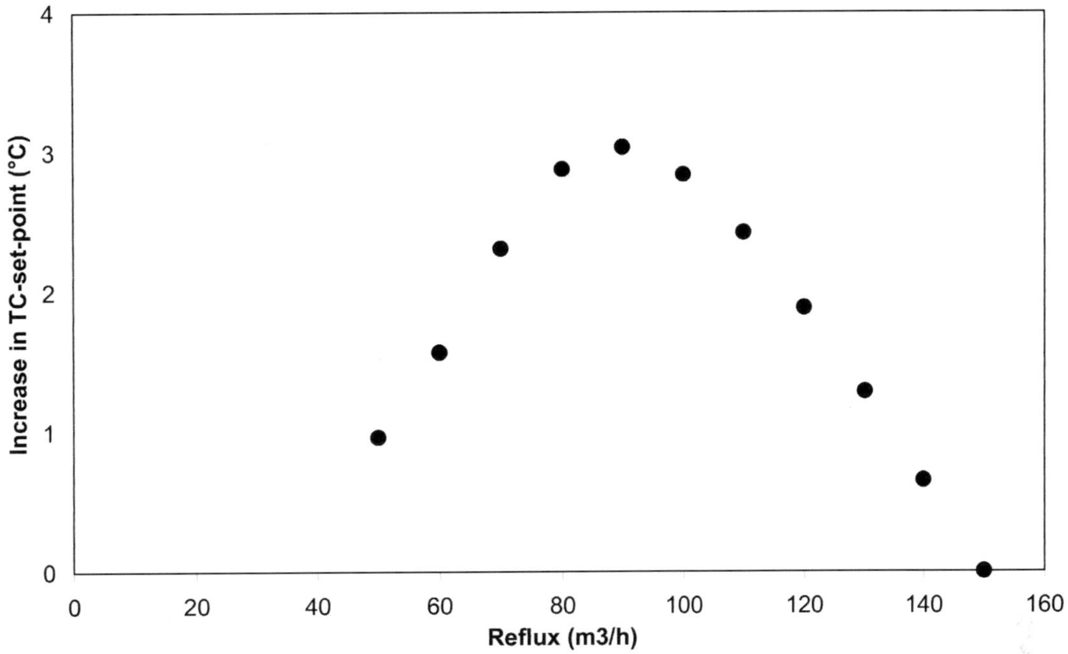

Figure 10. Simulated required increase in calculated control set-point for a constant concentration of 0.2 mole% i-pentane in the LPG.

3.6 Simulation results – condenser temperature

The overhead temperature has been fixed for a long time at about 35-36 °C. Decreasing this temperature to 25 °C for the base case would increase the LPG draw rate by more than 2 Sm^3/h, of which 0.4 Sm^3/h would be propane and almost 1.6 Sm^3/h would be butanes. This would increase product value and decrease the fuel gas production, which often limits the whole refinery throughput in summer. This potential needed to be verified experimentally.

3.7 Simulation results – lowering feed inlet temperature

As seen from Figure 5, the temperature on the theoretical tray just below the feed inlet is about 153.5 °C for the base case, whereas the temperature on the theoretical tray just above the feed inlet is about 130 °C. The inlet temperature has historically been about 150 °C. Lowering the inlet temperature from 150 °C to 140 °C reduces the feed pre-heat duty for the base case feed by 1.84 MW. This would reduce the reflux rate by about 30 Sm^3/h at constant i-pentane content in the LPG. n-butane in the naphtha would increase slightly, but with far less than for a reduction in reboiler duty by 1.84 MW. This means that there is a potential energy saving on the pre-heater. This potential needed experimental verification.

4 Test runs and evaluation of the controller

Several test runs were made, the purpose of which were to test the performance of the debutanizer under a set of predefined conditions with the revised temperature control scheme in place. We also wished to verify the simulation results regarding the optimum weighing coefficient and to establish the best working set-point for the new temperature controller. A secondary goal was to see if it was possible to save energy in the debutanizer's reboiler and/or feed pre-heater as indicated by the simulations.

A novelty since 2006 was that the naphtha stabiliser on the crude unit had been equipped with a calculation of the C_4 content in the stabilised naphtha and that there was a DMC controller dedicated to keep that concentration. The controller on the crude stabiliser worked well during steady flow rates, but did not work perfectly during crude changes or during changes in feed rate. Another difference in 2009 compared to 2006 was that the debutanizer top pressure had been reduced from 10.7 barg to 10.0 barg.

A drawback since 2006 was that the online analysers were no longer working properly and often gave misleading values.

4.1 Extremely low reflux

The controller was launched when the debutanizer had problems with flooding. At reflux + LPG draw rate exceeding about 100 Sm^3/h, the tower flooded. The problem started after an upstream water wash and was believed to be due to salt plugging up the top tray or top trays. The flooding problem was later solved by online water washing via the reflux line. In order to avoid flooding when the controller was taken in service, both pre-heating and reboiler duty were decreased. Tower inlet temperature was 145 °C and the reflux was 88 Sm^3/h at a naphtha feed rate of 304 Sm^3/h giving a reflux ratio (reflux rate/feed rate to HDS) of 0.29. The butane concentration in the naphtha from the crude stabilizer was about 2.4 volume% according to the online calculation. For a weighted control temperature of 160 °C the LPG contained 0.38 mole% i-pentane and 0.05 mole% n-pentane. The naphtha contained 0.09 volume% n-butane. Thus, the product purities were much higher than previously reported average values, even at this extremely low reflux ratio. This verified that the tower had been over reboiled to compensate for the poor cut-point control.

As the flooding problem was solved, we re-established normal operation before further evaluation of the debutanizer was made.

4.2 Low reflux test run

In the low reflux test run, reflux was kept at about 110 Sm^3/h and tower pre-heat was the same as had historically been used. The bottoms flow rate was just below 300 Sm^3/h (see Table 1) and the reflux ratio was about 0.38 (compared to 0.45-0.5 as before). The new controller was used and the set-point for the controller temperature was varied from 154 to 162 °C (Table 3).

The relationships between tower temperatures and top and bottom qualities are shown in Figures 11 and 12. In Figure 13 the corresponding concentrations of n-butane in the naphtha and i-pentane in the LPG are plotted. As one can see, the cut is not perfect, but far better product qualities than the old average values for both specifications could be attained simultaneously for control temperatures between 158 and 160 °C.

Figure 14 shows a similar temperature curve as was shown in Figures 4 and 8. The curvature is similar to Figure 4.

Table 3. Operational parameters during the low reflux test run. Average values during more or less steady state conditions.

TC set-point	154	156	158	160	162
Tower top pressure (barg)	10.0	10.0	10.0	10.0	10.0
Inlet temperature	151.0	151.0	151.0	151.1	151.0
OH-temperature	35.9	34.6	39.2	36.2	36.3
Top temperature	70.9	70.6	71.2	72.3	72.8
Temperature on tray # 8	81.5	81.8	82.7	85.3	97.3
Temperature on tray # 24	161.3	163.6	166.0	167.6	168.4
Bottom outlet temperature	193.6	194.0	194.6	195.0	195.6
LPG-draw (Sm^3/h)	7.0	7.1	6.6	8.3	9.1
Reflux (Sm^3/h)	111.3	109.8	110.4	110.6	111.3
Reflux ratio	0.38	0.37	0.38	0.38	0.38
Off-gas flow rate (Sm^3/h)	3288	3284	3797	3078	3155
Bottom flow rate (Sm^3/h)	298.4	299.0	296.6	298.8	298.7
n-butane in naphtha (vol%)	0.550	0.377	0.215	0.113	0.0628
i-butane in naphtha (vol%)	0.0114	0.0081	0.0056	0	0
i-pentane in LPG (mole%)	0.03	0.03	0.06	0.17	1.20
n-pentane in LPG (mole%)	0	0	0.01	0.02	0.17

Unfortunately, the correlation between LPG draw rate and tray temperature observed in the test data was not of sufficiently high quality to be used for fine tuning of the η coefficient. The correlation was disturbed by the variable amount of butane that was lost to off-gas. This loss was not constant due to inability to maintain constant temperature from the overheads condenser. Control of OH-temperature was made difficult due to atmospheric disturbances and by the control range of the variable speed fans.

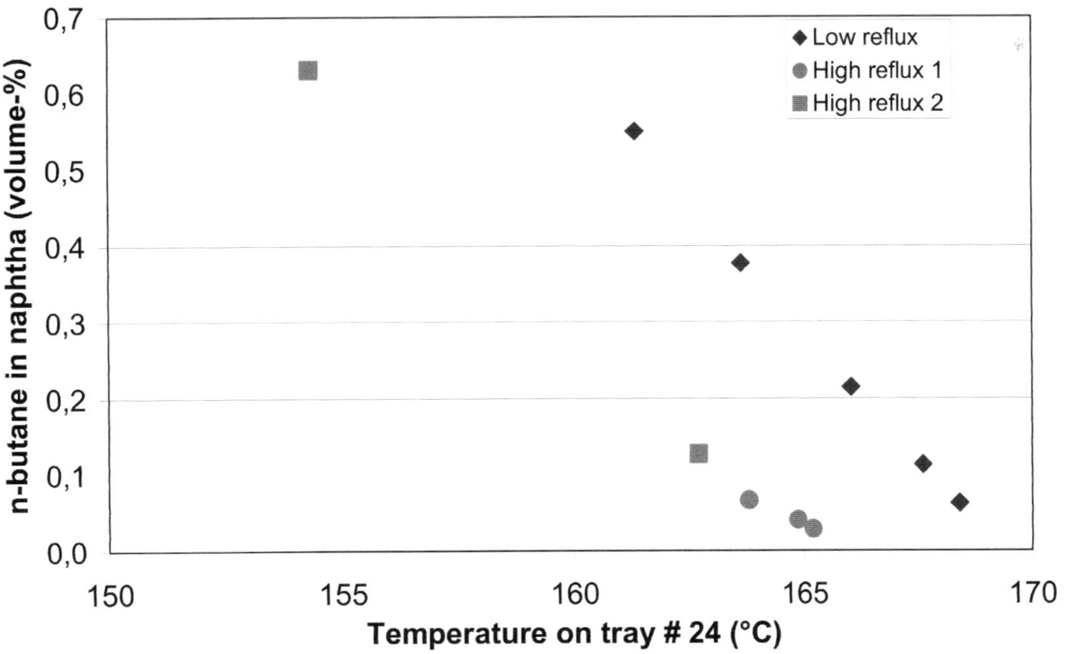

Figure 11. Correlations between concentration of n-butane in the bottoms naphtha and the temperature on tray # 24 for low and high reflux tests.

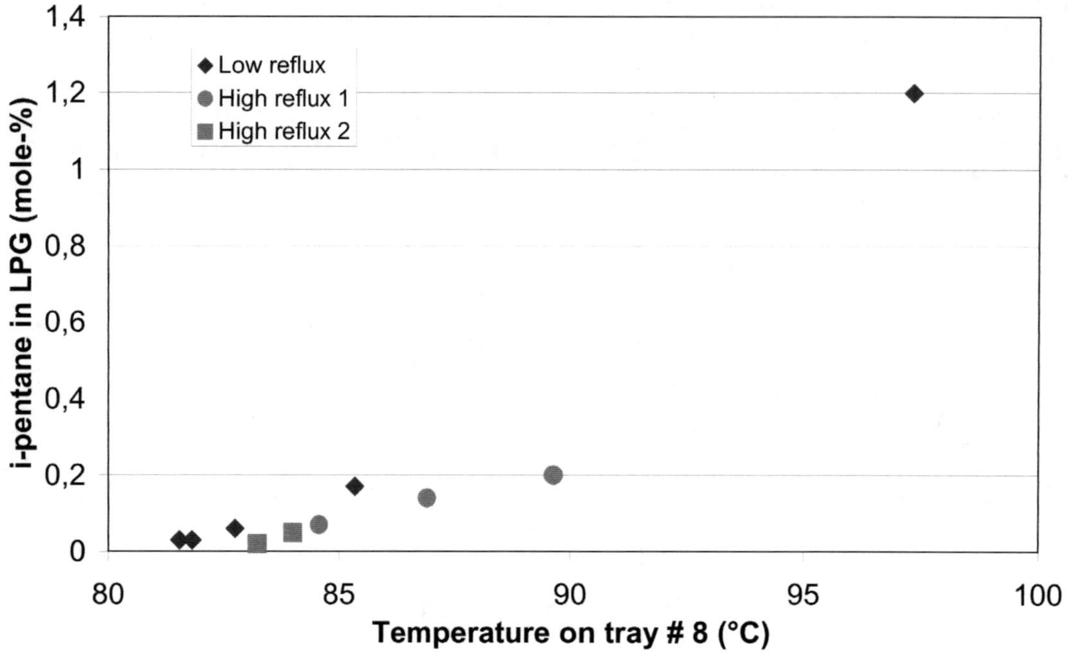

Figure 12. Correlations between concentration of i-pentane in the LPG and the temperature on tray # 8 for low and high reflux tests.

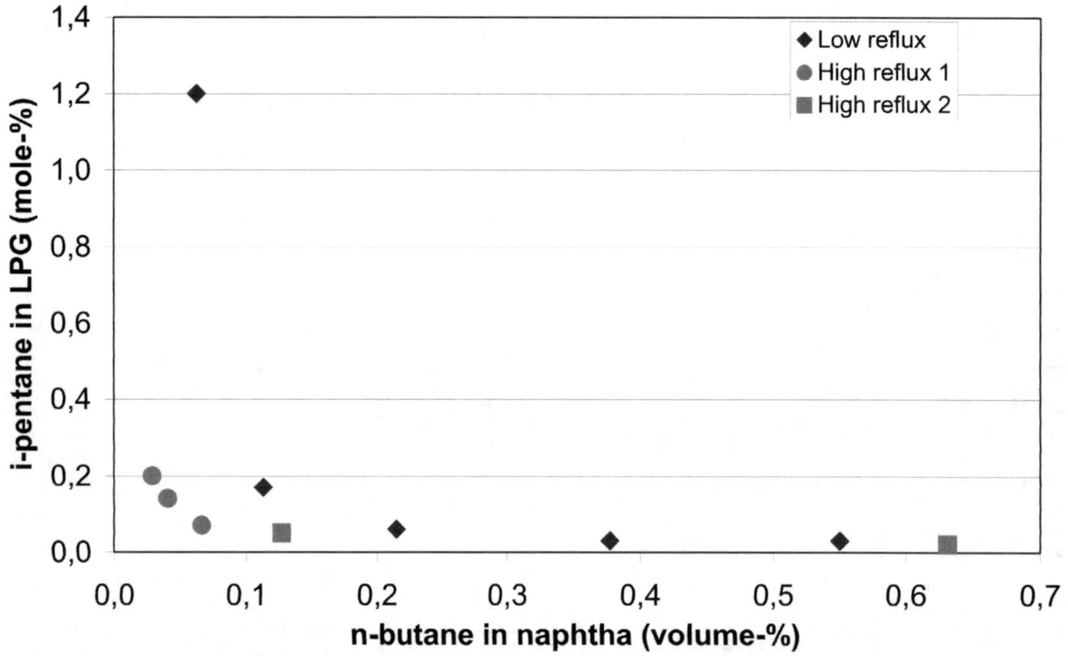

Figure 13. Quality of separation: Corresponding concentrations of n-butane in the naphtha and of i-pentane in the LPG for low and high reflux tests. Average values before the controller was taken in service were 1.0 mole% pentane in LPG and 0.33 volume% butanes in naphtha.

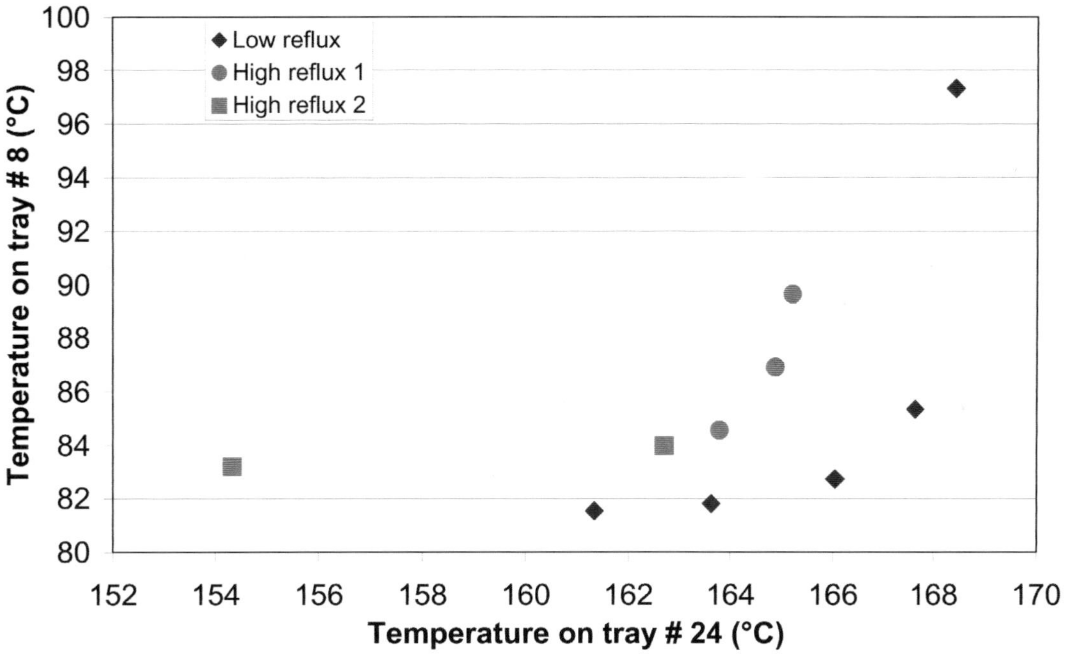

Figure 14. Corresponding temperatures on tray # 8 and on tray # 24 for low and high reflux tests.

4.3 High reflux test runs

During the high reflux test run, the reflux ratio was kept at about 0.5 (see Table 4). Three different control temperatures were tested (Table 4).

A complementary high reflux test was made at a later date in which every effort was made to minimise the impact of the primary disturbances identified in previous tests, feed composition

Table 4. Operational parameters during the high reflux test runs.

TC set-point	158	157.5	156.5	-	-
Tower top pressure (barg)	10.0	10.0	10.0	10.0	10.0
Inlet temperature	151.0	151.0	151.0	150.0	150.0
OH-temperature	36.2	36.2	35.9	33.1	33.2
Top temperature	72.2	71.9	71.4	70.6	70.2
Temperature on tray 8	89.6	86.9	84.6	84.0	83.2
Temperature on tray 24	165.2	164.9	163.8	162.7	154.3
Bottom outlet temperature	195.6	195.1	195.1	194.4	192.6
LPG-draw (Sm^3/h)	7.3	7.3	8.0	7.0	6.0
Reflux (Sm^3/h)	150.1	149.6	143.3	137.3	143.5
Reflux ratio	0.51	0.51	0.50	0.48	0.50
Off-gas flow rate (Sm^3/h)	3229	3327	3442	3268	3292
Bottom flow rate (Sm^3/h)	298.9	298.7	287.6	286.6	288.7
n-butane in naphtha (vol%)	0.0290	0.0409	0.0665	0.1270	0.6310
i-butane in naphtha (vol%)	0	0	0	0	0
i-pentane in LPG (mole%)	0.20	0.14	0.07	0.05	0.02
n-pentane in LPG (mole%)	0.02	0.02	0.01	0.01	0

and condenser temperature. At this time the feed rate to the naphtha unit was constant and the other units producing naphtha were running stably. This was considered a golden opportunity to vary the LPG draw rate manually to get an evaluation of whether or not the weighing factor was optimum. Great attention was paid to the OH-temperature.

As long as the draw rate was low enough for butanes to remain in the bottoms naphtha, the tower temperatures were constant within ±0.5-1 °C. The calculated butane concentration in the naphtha from the crude stabiliser varied between 1.9 and 2.1 volume-% for most of the test, but was disturbed by a change in feed rate and was within 1.8-2.2 volume-% when the LPG draw rate was at its highest. When the draw rate was increased so that all butane ended up in the LPG, the temperature on tray # 8 became too unsteady to draw any conclusion (this test point was therefore left out from Figures 11-14). This was probably due to the variations in butane in the feed and because the temperature on tray # 8 is so very sensitive to the draw rate as pentanes start to come over the top. We were unable to verify that a weighing factor of 0.1 is ideal, but it was confirmed that it should be small. The controller works satisfactorily so it was decided that the original weighing should be kept.

4.4 Disturbance in butane content in feed

As already mentioned, the most severe disturbance to the tower before the new controller was taken in service, was the variation in butane content in the naphtha feed that followed a crude switch. The new controller handled these step changes very effectively.

Figure 15 shows the response of the debutanizer when the n-butane in the naphtha from the crude naphtha stabilizer was increased rapidly from about 2.1 to 3.5 volume%. As more butane comes into the debutanizer, the temperature on tray # 24 starts to drop. The LPG draw rate is therefore increased by the controller and a new steady-state condition is soon established. The maximum drop in tray # 24 temperature before the temperature started to increase again due to the increased LPG draw rate was 1.8 °C. Hence, not much butane is lost in the bottoms due to the disturbance. The amount of n-butane lost in the naphtha when the temperature on tray # 24 drops can be seen in Figure 11.

The stabilised naphtha from the crude unit as well as the total naphtha charge to the naphtha hydrotreating unit was sampled before and after the step change in butane concentration in the naphtha from the crude stabilizer. Sampling of crude naphtha at “1.75 hours” as per Figure 15 gave a n-butane concentration of 2.13 volume% and 0.28 volume% i-butane from the crude unit and the naphtha feed to the naphtha hydrotreating unit contained 2.17 volume% n-butane and 0.35 volume% i-butane. Sampling at “24 hours” showed that the naphtha from the stabilizer contained 3.45 volume% (3.52 according to the online calculation) n-butane and 0.82 volume% i-butane, while the total naphtha feed to the naphtha hydrotreating unit contained 2.96 volume% n-butane and 0.66 volume% i-butane.

The increase in total butane content in the naphtha to the hydrotreating unit was 1.1 volume%. The total naphtha feed rate increased from 327 to 330 Sm3/h (it was attempted to keep the bottoms naphtha flow rate from the debutanizer constant). The rate of butanes in the naphtha feed to the naphtha unit increased by about 3.7 Sm^3/h. The increase in LPG draw rate was about 4.2 Sm^3/h between the two operational cases. The off-gas flow rate decreased as more butane came into the naphtha unit.

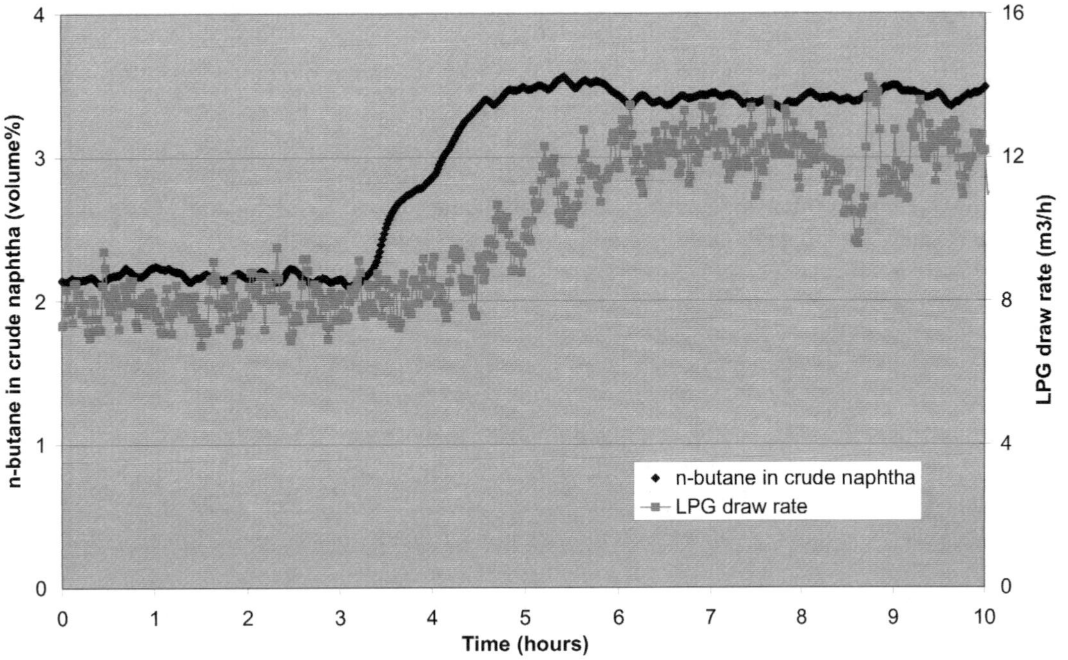

Figure 15. The response of LPG draw rate to an increase in n-butane content in the naphtha from the crude unit's stabiliser. The content of n-butane in the naphtha to the hydrotreating unit increased from about 2.2 volume% to about 3.0 volume%.

4.5 Change in naphtha feed rate

Variations in naphtha feed rate is the one disturbance the controller has problems with. As the online analysers have not been working properly for most of the time since the controller was taken in service, the operators have not had a reliable guide for the controller temperature set-point. To keep the i-pentane concentration in the LPG at 0.2 mole%, a TC set-point of 158 °C is required at a reflux ratio of 0.51 (Table 4), whereas at a reflux ratio of 0.38, a TC set-point slightly above 160 °C is needed for the same concentration of i-pentane in the LPG (see Table 3). These reflux ratios correspond to reflux rates of about 150 and 110 Sm^3/h for the simulation base case feed rate. As seen from Figure 10, a difference in TC set-point of about 2.4 °C would be required to keep the same concentration of i-pentane in the LPG according to the simulations and this is verified by the test runs.

With well functioning online analysers, the operators could easily handle changes in feed rate. Without proper guidance, some help to the operators was required. Two options were tested. The first was to save energy and to focus on the LPG specification and sacrifice butanes in naphtha. This option was based on the observation that the debutanizer was over reboiled. Adequate separation is achievable with the dedicated reboiler shut-off. In this condition, at least one product purity, and for most of the time both top and bottom product purities were better with the new controller. By setting the TC set-point to 157 °C, the top product is always on specification for all reflux ratios between 0.25 and 0.5 (as long as feed inlet temperature remained at 150 °C). This option is preferred when a fuel gas deficiency exists at the refinery and is further discussed in Section 5.2.

The second option that was tested was to simply remove the disturbance by controlling the reflux ratio. This requires that the dedicated reboiler is left in service. Since both product purities can be kept much lower than before the controller, even at reduced separation, there is room for lowering the average reboiler duty. However, there is a limit to how much the reboiler duty can be lowered, since the heater must be within its control range for all process conditions. For this option, it was considered best to lower the tower inlet temperature by 10 °C and to lower the reboiler duty slightly. The flow of fuel gas to the dedicated reboiler was controlled to maintain the reflux ratio. This is further described in section 5.1.

4.6 Reduction in pre-heat

The inlet temperature was decreased from 150 to 140 °C. This corresponds to a decrease in pre-heat duty of about 2 MW when the naphtha product flow rate is about 320 Sm^3/h. This energy was picked up in a downstream kettle type reboiler for another tower. Since that tower has an internal make-up reboiler using steam as heating medium, the internal reboiler could be unloaded by 2 MW which corresponds to the condensation duty of about 3.5 tons of steam/h.

The 10 °C decrease in feed inlet temperature led to a decrease in reflux ratio of about 0.06. Using a reboiler duty between what we used in the previous “low reflux test” and “high reflux test”, gave a reflux ratio of 0.36 with the inlet temperature at 140 °C. Table 5 shows the product qualities when the TC set-point was 157.5 °C.

Table 5. Three different test conditions with debutanizer inlet temperature decreased to 140 °C.

TC set-point	157.5	157.5	156.5
Tower top pressure (barg)	10.0	10.0	10.0
Inlet temperature	140	140	139
n-butane in feed to HDS reactors (vol%)	-	2.17	2.96
i-butane in feed to HDS reactors (vol%)	-	0.35	0.66
Reflux ratio	0.36	0.30	0.30
n-butane in naphtha (vol%)	0.0464	0.1592	0.1546
i-butane in naphtha (vol%)	0	0	0
i-pentane in LPG (mole%)	0.28	0.10	0.30
n-pentane in LPG (mole%)	0.03	0.01	0.03

Separation was still good, although the TC set-point was on the high side. The TC set-point was therefore lowered to 156.5 °C for the following tests. First, reboil was reduced and controlled by the reflux ratio controller to keep a constant reflux ratio of 0.30. Separation was still good (Table 5).

Later, the reboiler duty for the crude stabilizer was decreased and more butane was left in the stabilizer bottoms which caused the amount of butane in the feed to the debutanizer to increase. The resulting product purities are shown in Table 5. Separation was still much better than before the controller was taken in service and the energy consumption for a naphtha charge rate of 330 Sm^3/h had been decreased by 2 MW in the pre-heater and by more than 2 MW in the reboiler heater.

4.7 Evaluation of the test runs

In Figure 13 the corresponding concentrations of n-butane in the naphtha and of i-pentane in the LPG are plotted. The diagram shows clearly that high reflux results in lower concentrations of n-butane in the naphtha at the same levels of i-pentane in the LPG. This confirms that better separation can be achieved at higher reboiler duties/reflux rates.

The higher the reflux ratio, the better separation one has and for a fix butane concentration in the naphtha, the higher concentration of butane one has on tray # 24 and hence the lower tray temperature one has on tray # 24. This is seen in Figure 11. The better the separation, the higher pentane concentration there is on tray # 8 for a fixed pentane concentration over top in the debutanizer and the higher tray temperature one has on tray # 8. This is seen in Figure 12. These facts complicate composition control slightly, since there is not a single temperature on tray # 8, nor a single calculated control temperature that gives the same pentane concentration in the LPG (or the same butane concentration in the naphtha) for all reflux ratios. The greater the reflux ratio one uses, the lower control temperature is required to get the same pentane concentration in the LPG.

In order to address the question as to what product qualities to aim for one must consider the overall refinery economics. During summer time, the amount of fuel gas sometimes becomes so high that it limits the whole refinery capacity (flaring is forbidden by law). At such times it is beneficial to increase the fuel gas consumption (increased reboiler duty and increased separation) and minimise slippage of butane to the fuel gas system (low butane concentration in the bottoms naphtha). During winter time, there is often a shortage of fuel gas and either propane or butane products are vaporised and sent to the fuel gas system. During such times, it is better to minimise fuel gas in the reboiler and to pay less attention to the butane concentration in the bottoms naphtha.

The new controller gives better quality products and potential energy savings. Either the original reboiler duty and pre-heat duty can be retained with product impurities as low as 0.07 mole-% i-pentane in LPG and 0.07 volume-% n-butane in the naphtha (see Table 4) compared to 1 mole-% and 0.33 volume-% before the controller. Or the reflux can be reduced by 30-40% by reducing the reboiler duty and both specifications can be better than before (average values) or one of the specifications slightly better than before (average values) while keeping the other specification at much lower levels than before.

5 Summary of new operating modes

5.1 Control of reboiler duty cuts energy consumption and makes cut-point control easier

As was seen in the test runs and in the simulation, the ideal calculated control temperature depends on the reflux ratio. Since acceptable fractionation can be achieved using a lower than normal reflux ratio, there is little incentive to use a high reboiler duty, unless fuel gas is in excess.

One of the modes of operation employed with success for some time involved a permanent decrease in feed pre-heat duty by decreasing the tower feed inlet temperature by 10 °C and an on average decrease in reboiler duty. The reboiler duty was controlled as to keep a constant reflux ratio. This control scheme eliminated the only disturbance that requires updates in control temperature set-point. The reflux ratio was chosen as to keep the reboiler in its operating range while keeping the average duty as low as possible.

With the reflux ratio controller in operation, the ideal weighted control temperature was constant as long as the reflux ratio set-point and/or inlet temperature was not changed by the control room operator. When the tower feed temperature was 140 °C and the reflux ratio was 0.30, the concentration of i-pentane in the LPG was 0.1 mole-% while the concentration of n-butane in the naphtha was about 0.16 volume% when the control temperature set-point was 156.5 °C (Table 5).

5.2 Shut-off reboiler heater saves fuel gas while the LPG purity is still controlled

An alternative mode of operation that has been tested was to simply shut off the dedicated make-up reboiler. Since the average fuel gas consumption in this furnace used to be about 500 Nm^3/h of fuel gas, this corresponds to quite an energy saving. In order to maintain separation, the tower inlet temperature was kept at its original value of 150 °C. For this mode of operation, the reflux ratio varies with time. In the absence of reliable online instruments, we decided to focus on the LPG specification which is why the weighted control temperature was kept at 157 °C regardless of reflux ratio.

6 Summary of improvements

In summary, the control scheme based on weighted temperature works very well. We have found two different modes of operation, one that saves fuel gas and controls LPG purity and one mode of operation that gives a lesser saving in fuel gas, but saves steam consumption and that allows for simultaneous control of both LPG purity and naphtha purity. Both modes of operation have unloaded the tower overhead condenser, permitting operation at a lower condenser temperature than before. This lower OH-temperature increased the LPG yield (of the order several Sm^3/h when OH-temperature is decreased to 25 °C) at the expense of off-gas.

Table 6. Different modes of operation and their advantages.

Case	Temperature controller	Feed rate Sm^3/h	Reflux ratio	Inlet temperature	n-C_4 in naphtha feed volume%	C_5 in LPG mole%	C_4 in product naphtha volume%	Improvement compared to base case
Original (average)	No	280-350	0.45-0.50	About 150	Probably around 2	1	0.33	Non (Base case)
Controlled	Yes	300	0.51	151	2	0.08	0.07	Product purity
Reduced reboil	Yes	290	0.38	151	2	0.07	0.22	Purity +fuel gas saving
Reduced pre-heat	Yes	343	0.36	140	2	0.32	0.046	Purity +steam saving
Reduced pre-heat and reboil	Yes	330	0.30	140	2.2	0.11	0.16	Purity + steam and fuel gas saving
Reduced pre-heat, reboil and increased C_4 in feed	Yes	330	0.30	139	3.0	0.33	0.15	Purity + steam and fuel gas saving + energy saving in crude unit

Both the concentration of i-pentane in the LPG and of n-butane in the bottom naphtha are controlled very well and both concentrations are far lower than the previous average concentrations, even though the total energy usage has been significantly reduced. Table 6 lists the base case (normal/average operation before the controller) and some different modes of operation that has been tested, together with the advantages with the new modes compared to the base case.

7 Acknowledgements

The authors wish to acknowledge the laboratory technicians at Preem refinery in Lysekil, in particular Annette Antonsson and Cecilia Dahlström-Patriksson, for their efforts in providing us with a large quantity of information of high quality.

8 Reference

1. "Evaluation of Criteria for Selecting Temperature Control Trays in Distillation Columns", Luyben,W. L., Journal of Process Control 16 (2006), p 115-134.

(102f) Gamma Scanning Fundamentals - Confirmed!

Lowell Pless, Tracerco, Pasadena, TX and Izak Nieuwoudt, Koch-Glitsch, Wichita, KS

Gamma Scanning Fundamentals – Confirmed!

One of the most common process troubleshooting applications, especially when it comes to separation and fractionation equipment is gamma scanning. Gamma scans provide a density profile of the internal process of operating distillation columns and other process vessels. The density profile can be used to diagnose the hydraulic operating conditions of mass transfer devices such as damage to internals, flooding, degree of entrainment or weeping, liquid levels on trays and distributors, liquid distribution through packed beds, etc.

There were research and development experiments with gamma scanning back during its development years in the 1960's and 1970's. One scanning service provider, Tru-Tec, performed some tests using a glass column at Koch Research facilities in the mid 1980's. But since that time it appears no basic development or confirmation work has been done. During these intervening years radiation detector technology has been improved.

Tracerco performed a series of gamma scans with the cooperation of Koch-Glitsch on one of their pilot plant towers equipped with trays. A series of scans was done under varying operating conditions in order to correlate scan results with verified process conditions. The results learned, or re-learned, will be presented here.

See more of this Session: Contributions to Distillation Troubleshooting and Control
See more of this Group/Topical: Topical 8: The Dr. James Fair Heritage Distillation Symposium

Thermal Integration of Re-boilers

E.E. Vazquez –Ramirez , M. Riesco Avila, D. Jantes Jaramillo &
G.T.Polley
University of Guanajuato, Mexico

The approaches currently used by process integration specialists to assess the opportunities for the integration of re-boilers are frequently erroneous and lead to proposals that have to be rejected at the detailed engineering stages (or worse, to re-boilers w commissioned do not perform properly).

This paper examines the integration of both vertical and horizontal thermo-siphon reboilers.

It covers the following topics: how should the heat load-temperature characteristics be incorporated into a "pinch analysis", how should the hot stream be matched with the re-boiler duty, what is the minimum temperature difference for the re-boiler, what is the maximum temperature difference for the re-boiler, how does distillation column control affect the integration and when should indirect integration be employed.

Heat Load – Temperature Characteristics

Consider the vaporisation of a binary mixture. If we ignore (for the moment) the temperature difference required to initiate vapour generation, the vaporisation of a liquid mixture commences once the "bubble point" of the mixture is exceeded. As vapour is generated the concentration of the less volatile component increases and this leads to an increase in saturation temperature. If the design of the re-boiler is such that liquid and vapour are in contact throughout the vaporisation process (achieved in vertical tubes) then the vaporisation process could continue until all of the liquid is evaporated at which point the saturation temperature has reached the "dew point" of the mixture.

If the design of the re-boiler is such that liquid and vapour flows are separated (as occurs in horizontal tube bundles, see Vazquez –Ramirez et al, 2011) then vaporisation is completed at the boiling point of the less volatile component.

These are limiting values. In practice, even in once through designs, the feed to a re-boiler does not equate with the quantity of material that is to be vaporised. With a vertical thermo-siphon re-boiler the maximum amount will rarely exceed 20% of the feed and in most applications will be less than 10 % of the feed. Horizontal thermo-siphons the fraction of feed vaporised should be lower.

A reasonable temperature-heat load relationship for use in evaluating how a re-boiler can be integrated into a process would a heat load equally the required boil-up absorbed at a temperature calculated for 10% vaporisation of the mixture being evaporated. This temperature is determined from the heat demand curve for the mixture:

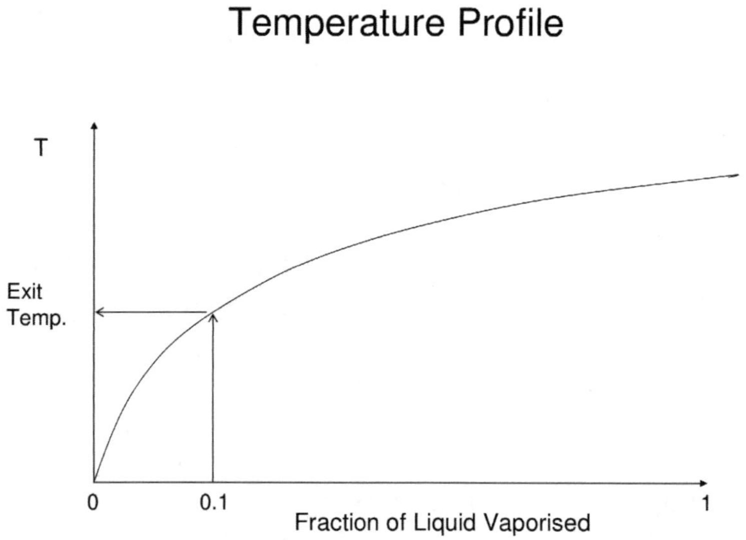

Figure 1. Temperature at re-boiler exit for 10% exit quality

Matching of Streams Within a Re-boiler

If the stream being used to provide heat to the re-boiler is a liquid stream then it should its flow should be co-current with flow of the stream being vaporised. This is particularly important with vertical thermo-siphon re-boilers.

The mixture being vaporised flows through the tubes of a vertical thermo-siphon. It enters the tubes as a sub-cooled liquid (the liquid in the reservoir will be at saturation temperature but the surface of the reservoir can be 3-4 metres above the lower tube-sheet of the re-boiler). The point at which nucleation (the formation of bubbles of vapour) first occurs within the tube is dependant upon both the temperature of the liquid and the temperature of the hot wall. Maximising temperature driving force reduces the length of tube prior to the onset of nucleate boiling. This driving force is maximised if the hot liquid flows co-currently.

The heat transfer coefficient on the cold-side of the unit increases rapidly once a two-phase flow has been established. Consequently, there is a large variation of heat transfer coefficient along the length of the tube. This variation is "stabilised" if co-current flow is used. If the hot liquid flows counter-current to vaporising mixture its temperature profile is affected by changes in the heat transfer distribution on the cold-side. These changes affect the point at which nucleation commences and the heat transfer distribution on the cold-side. So, the use of counter-current flow can result in instability and control difficulties.

Incorporating Re-boiler Characteristics in the Cold Composite Curve

The concept of "composite curves" is based upon the assumption of counter-current flow. Co-current flow should be used in a thermo-siphon re-boiler. So, how can such heat demand be represented in a composite curve? The option recommended here is that it is represented as an "isothermal" operation being conducted at the vapour exit temperature (Figure 2).

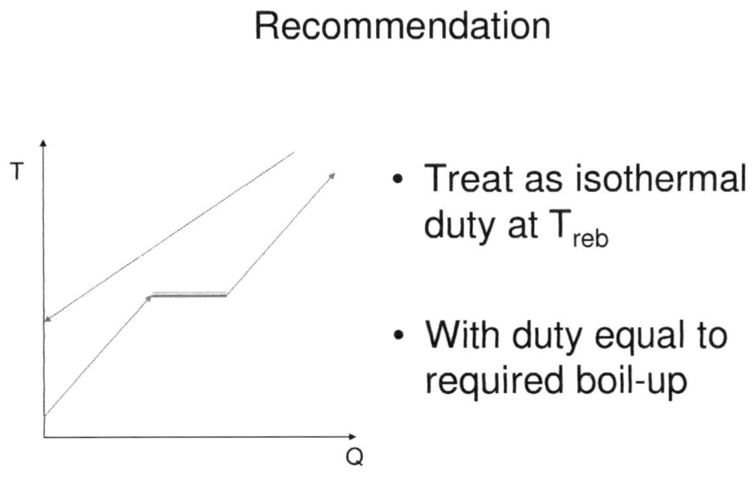

Figure 2. Incorporating Vaporisation Duty into Cold Composite Curve

Allowing for Start-up and Control

Start-up and control are important considerations that must made when evaluating re-boiler integration.

Whilst a plant may be designed to operate at a given condition (e.g. specific throughput) it will be operated under a range of conditions. This means that re-boiler load will be subject to variation and control will be required.

In many situations it will be found that start-up and control cannot be undertaken using an integrated unit and that the overall duty must be undertaken using two separate re-boilers: an integrated unit having a fixed duty and an on-integrated unit used for both start-up and for control function.

The engineer needs to determine the following:

1. the load required for start-up purposes (this is set by the minimum loading on the column not the ultimate operating condition);
2. the variation in load required for the control function

The design of the non-integrated re-boiler will involve identifying a geometry that will provide either the start-up performance or the load associated with maximum control load (whichever is the larger) at a point that is at least twenty percent below the maximum operating condition for the design (set by critical heat flux conditions).

1. This re-boiler needs to be on-stream all of the time the plant is operational. Consequently, its load will never be zero. The minimum value should be set at a value that is around twenty percent above the lowest stable condition (the minimal sustainable re-boiler load);

The load for the integrated unit will be the total load minus that required at the minimum throughput minus that needed by the non-integrated unit operating at its minimum load.

Minimum Temperature Approach

Whilst the re-boiler is represented by an isothermal line set at the temperature at which the boiling mixture exits the re-boiler checks do need to be made that the temperature of the hot stream entering the re-boiler is sufficient for it to operate.

With units handling aqueous mixtures the minimum wall temperature that has to be achieved within the unit should be at least 15 C above the saturation temperature of the liquid entering the unit (note that in vertical thermo-siphon re-boilers this saturation temperature is higher than that within the reservoir feeding the unit – in the case of units operating under vacuum the difference can be large). For units handling organics the required difference between these temperatures can be taken as 5 C.

Maximum Temperature Approach

In some applications there is a limitation on the maximum temperature of hot fluid used to drive the re-boiler. This is set by the rate at which vapour is generated close to entry.

Kettle Re-Boilers

With a kettle re-boiler liquid enters the tube bundle from the base and sides (Figure 3).

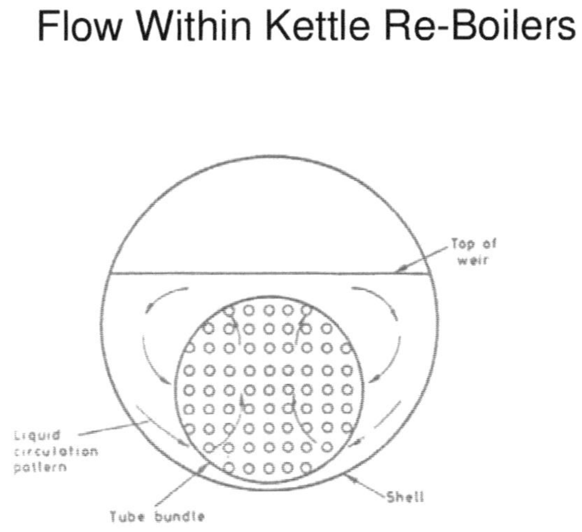

Figure 3. Operation of a kettle re-boiler

The tubes at the periphery of the bundle are well wetted. However, as the liquid endeavours to move into the centre of the bundle it encounters an increasing quantity of vapour. The result can be difficulty in wetting the tubes within the core of the bundle.

This gives rise to a relationship between temperature driving force and heat flux that is very different to that observed for a single tube (Figure 4). The "plateau" observed with tube bundles is due to a situation in which the heat transfer coefficients for the tubes at the periphery are increasing due to increased nucleate boiling whilst those for the tubes within the core are falling due to liquid starvation.

This has lead to a design philosophy based upon a "limiting heat flux". This philosophy can easily be achieved when steam is used to drive the re-boiler (for the pressure within the steam chest reduces until the controlled load is obtained). However, where a hot liquid is to be used this is much harder to achieve for the actual heat flux is controlled by the inlet temperature of the liquid and the flow conditions on the tube-side of the unit.

Heat Transfer Performance

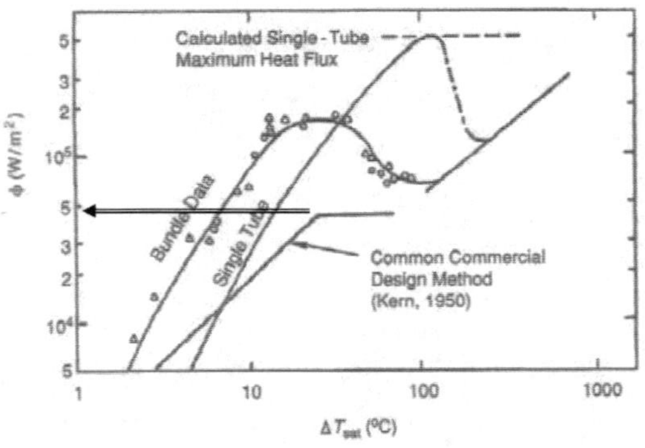

Design on basis of a maximum heat flux.

Figure 4. Thermal behaviour of Kettle Re-Boilers

Our conclusion is that this type of unit is best avoided when hot liquid is to be used as the heat source.

Vertical Thermo-siphon Re-boilers

One potential problem with vertical thermo-siphon re-boilers (Figure 5) operating with hot liquid as a heat source is excessive generation of vapour in the lower parts of the tube. This can give rise to a "choking" effect where such much vapour is generated there is insufficient head provided in the down pipe feeding the unit to drive the two-phase flow through the unit.

The result is vapour travelling back through the feed leg. Operation of the unit becomes unstable. Tubes dry-out and fouling or erosion/corrosion can occur.

This problem is most commonly encountered when the ratio of the liquid to vapour density is high (found when vaporising organics of molecular weight less than 200 at atmospheric pressure).

The phenomenon is sensitive to re-boiler tube length (the longer the tube the more likely the problem).

Any limit on the temperature of the hot stream should be determined through analysis of re-boiler behaviour ahead of integration analysis.

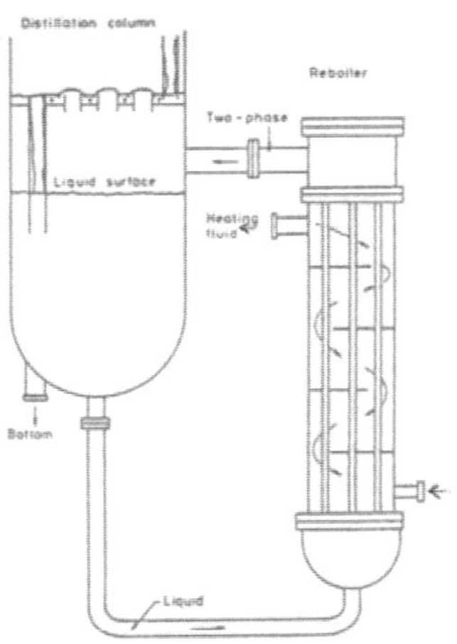

Figure 5. Vertical Thermo-siphon Re-boiler

Horizontal Thermo-siphon Re-Boilers

The rate of vapour generation is also a consideration that has to be made when design of horizontal units for use in integrated schemes is considered.

The generation of a lot of vapour in the first baffle spaces of a unit can give rise to phase mal-distribution problems further along the unit. Curves which allow the engineer to determine these effects have been presented by Vazquez-ramirez et al (2011) and are re-produced here.

Again the problem is more acute for duties involving high liquid-to-vapour density ratios.

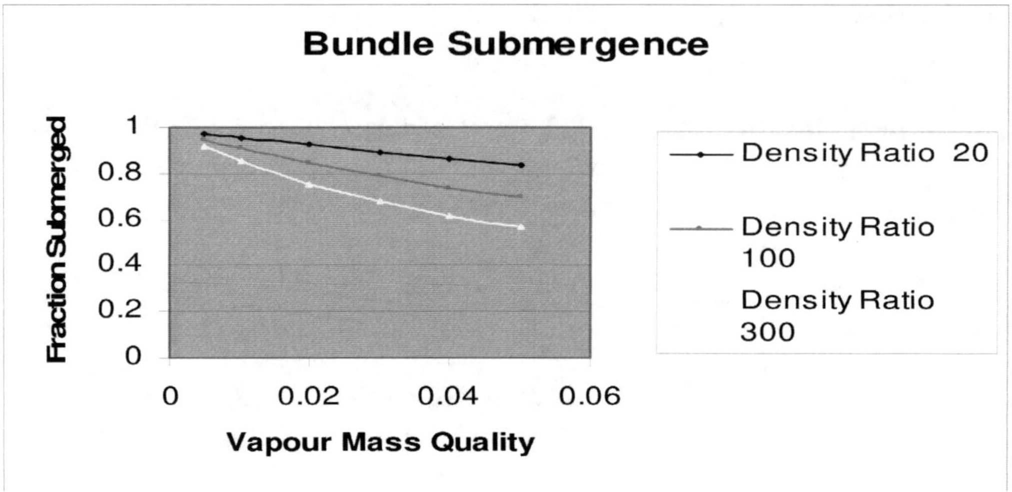

Figure 6. Side-to-side Flow

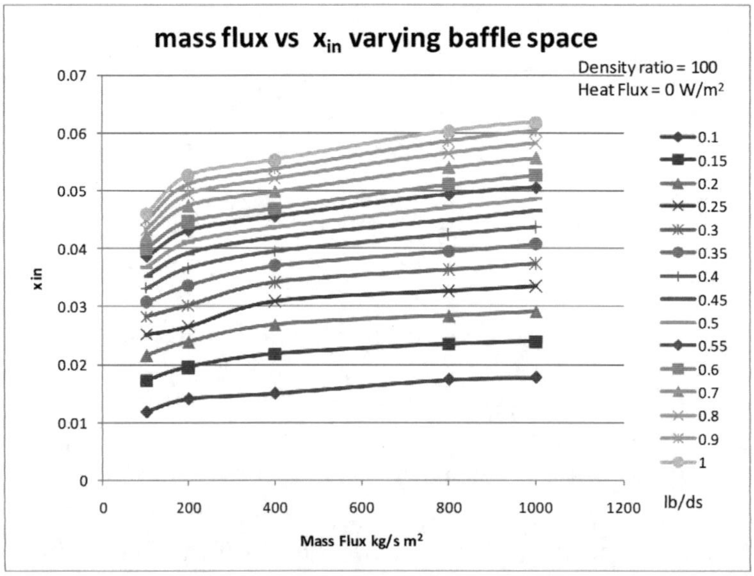

Figure 7. Up-and-down flow (Density Ratio of 100)

Conclusions

In this paper we have reviewed factors that need to be considered when integration of re-boilers is being evaluated.

Considered how re-boilers should be represented on composite curves and our recommendation differs from what we believe to be current practice.

We have looked at how a lower limit on temperature approach can be set. We do not believe that this is part of current practice.

We have discussed the need to establish an upper limit on temperature difference. This again is not part of current practice.

References

Vazquez –Ramirez E.E., Riesco Avila M. & Polley G.T. "Two Phase Flow through Horizontal Tube Bundles: Implications for Heat Exchanger Design", AIChE Spring Meeting, 2011

(107a) Designing Packed Multi-Partition Wall Dividing Wall Columns

Igor Dejanovic[1], Ljubica Matijasevic[1], Ivar J. Halvorsen[2], Sigurd Skogestad[3], Helmut Jansen[4], Bjoern Kaibel[4] and Zarko Olujic[5], (1)Chemical Engineering, University of Zagreb, Zagreb, Croatia, (2)Ict, SINTEF, Trondheim, Norway, (3)Department of Chemical Engineering, Norwegian University of Science and Technology, Trondheim, Norway, (4)Julius Montz GmbH, Hilden, Germany, (5)Process & Energy, Delft University of Technology, Delft, Netherlands

As proven in industrial practice, application of dividing wall columns (DWC), where appropriate, leads to approximately 30 per cent saving in both energy and capital, compared to equivalent conventional distillation sequences for obtaining three high purity products. Simulation studies performed with well defined, mainly hypothetical pure component systems, have indicated that designing four and more products DWCs could enhance potential energy and capital savings significantly. This implies that there is a strong incentive to consider implementing more complex designs of a DWC. These can certainly be assembled without great difficulties as structured packing columns, by utilizing well established non-welded, self-fixing partition wall technology that provides full flexibility in this respect. Nevertheless, the lack of experience and insight related to modeling and simulation of industrially relevant separations of multicomponent feeds, as well as suitable design procedures, accompanied often by a fear of unstable operation are main barriers in this respect.

The design and control of four-product DWCs have been addressed and encouraging advancements made in previous efforts, indicating that a packed DWCs could be arranged as a practical, single partition wall column, or a thermodynamically optimal, but complex, three-partition walls column. These and other simulation studies performed using so called V_{min}-diagram method as a practical means to identify and quantify accordingly internal configuration of a DWC, providing reliable data for initialization and effective execution of detailed calculations performed using an appropriate rigorous model arranged for this purpose using facilities available in a commercial software package, indicated a large potential for energy and capital saving by implementing DWCs in aromatics processing plants as encountered in complex refineries. Interestingly, it appeared that depending on the choice and specification of interesting products (components and/or fractions), a three-product, a four-product and even a five-product DWC could be an interesting option for aromatics processing plants.

Rigorous simulations provided optimum solutions with respect to stage and reflux requirements of alternative designs, and, most importantly imposed vapor splits that need to be realized in practice to enable proper operation of a complex DWC. The objective of this paper is to show how to dimension a multi-partition wall DWCs, i.e. how to arrange pressure drop in parallel sections, using available means (structured packings with different specific areas, corrugation inclination angles and corrugation designs in combination with state of the arts narrow trough distributors, and chevron and chimney tray type liquid collectors) in conjunction with appropriate predictive methods (available in public domain) to ensure achieving required vapor splits. The resulting column dimensions and related details in conjunction with adopted cost estimation methods allow a proper assessment of usability of a DWC. Our studies indicate clearly that a multi-partition wall DWC is an industrially viable option, which, if implemented, could contribute to a significant increase in profitability of aromatics processing plants.

See more of this Session: Dr. Fair's "Enhanced Distillation": Advances in Thermally Coupled and Reactive Distillation
See more of this Group/Topical: Topical 8: The Dr. James Fair Heritage Distillation Symposium

(107b) Design and Analysis of Multiple Thermally Coupled Configurations Using Column Profile Maps

Daniel A. Beneke, Ronald Abbas, **Diane Hildebrandt** and David Glasser, Center of Material and Process Synthesis, University of the Witwatersrand, Johannesburg, South Africa, Johannesburg, South Africa

This presentation introduces a method for rapidly assessing the feasibility of multiple thermally coupled separation units. The Column Profile Map technique has been applied to the design of a variety of quaternary feeds for a column consisting of a main column with various combinations of side rectifiers and/or strippers, as well as a fully thermally coupled column. The graphical nature of the technique allows one to easily assess feasible designs for systems with less than four components, but may be algebraically extended to higher-order systems using an eigenvector technique. Iso-reflux plots are presented as a tool to evaluate the minimum operating conditions of a specific column that also shows part of an Attainable Region and places bounds on operability. The fully thermally coupled arrangement is shown to require the least amount of heat addition for a variety of feeds, but also has by far the lowest thermodynamic efficiencies. We have further found that, subject to feed compositions, multiple side strippers, in agreement with modern crude refinery practice, have good first law expenditures with efficiencies comparable to other thermally coupled structures.

See more of this Session: Dr. Fair's "Enhanced Distillation": Advances in Thermally Coupled and Reactive Distillation
See more of this Group/Topical: Topical 8: The Dr. James Fair Heritage Distillation Symposium

(107c) Aspen Plus as a Tool for Process Scale-up of Externally Heat-Integrated Reactive Distillation

John C. Prindle[1], Chandrakant B. Panchal[2] and Richard D. Doctor[2], (1)Chemical & Biomolecular Engineering, Tulane University, New Orleans, LA, (2)Energy System Division, Argonne National Laboratory, Argonne, IL

Since its application to the manufacture of esters in the 1920s, reactive distillation has been recognized as a valuable unit operation. The simultaneous reaction and separation within a single piece of equipment can provide a lower capital alternative for those equilibrium reactions where the volatility of the products is substantially different. Higher conversions can often be achieved with reactive distillation than other process alternatives. Reactive distillation processes have been developed for both heterogeneous and homogeneous reactions – demonstrating the wide range of applicability this unit operation possesses. There are, however, several additional considerations during the development and scale-up of such a process. Degrees of freedom are lost when reaction and separation are combined in this manner. In addition, several of the relationships between distillation parameters (e.g. reflux ratio and distillate purity) do not hold universally for a reactive distillation system.

Additional degrees of freedom can be obtained through effective external heat integration and thermal management resulting in an optimum column temperature profile. This presentation will discuss the feasibility of several heat integration and thermal management techniques through process simulation with Aspen Plus. The potentially commercial two-column reactive distillation process for manufacturing cyclohexanol from cyclohexene via esterification and subsequent hydration will be used to illustrate the value of each technique. Other considerations during the development and scale-up of a reactive distillation process will be discussed as well.

See more of this Session: Dr. Fair's "Enhanced Distillation": Advances in Thermally Coupled and Reactive Distillation
See more of this Group/Topical: Topical 8: The Dr. James Fair Heritage Distillation Symposium

(107d) Reactive Distillation as a Multipurpose Reactor for Controlling Selectivity In Multiple Reaction Systems

Tobias Keller and Andrzej Górak, Laboratory of Fluid Separations, TU Dortmund, Dortmund, Germany

Reactive distillation (RD), in which the reaction and separation are integrated in a single unit operation, has been established in process technology. The instantaneous product removal from the reaction zone through distillation leads to an increased yield of chemical equilibrium-limited reactions, such as etherification and esterification. The majority of published papers on RD investigate chemical systems comprising only one main reaction. These papers have demonstrated the possibility of RD to increase reactant conversion as well as the purity of the desired product. In contrast, selectivity plays a secondary role in such chemical systems. In our opinion, the potential of RD has not been fully tapped. Our work focuses on multiple reaction systems, from which more than a single desired product is formed. We want to prove that RD enables the process engineer to flexibly vary the selectivity of each product while still having high reactant conversions in such chemical systems. The reactive distillation column can act as a multipurpose reactor, which consequently extends the current field of application for RD. No reports on the simultaneous control of conversion and selectivity in such a system have been published so far, and no experimental results on a pilot-scale for such a chemical system could be found.

A promising example is the reversible consecutive second-order transesterification of dimethyl carbonate (DMC) with ethanol via ethyl methyl carbonate (EMC) to produce diethyl carbonate (DEC). The transesterification of DMC was homogeneously catalysed by sodium ethoxide, and the reactive distillation experiments were performed in a pilot-scale RD-column with a diameter of 0.05 m that was equipped with Sulzer BX packing elements. The total packing height was 5.4 m, and the total feed rate was 4 kg/h. As temperature and concentration along the RD-column are measured, the experimental setup provides a detailed composition profile. The experiments demonstrated the potential of RD to act as a multipurpose reactor for multiple reaction systems producing more than one desired product. In the same column, selectivity for EMC higher than 80% and, after changing the operating conditions, selectivity for DEC higher than 80% could be achieved.

Additionally, the experiments provided reliable data for the validation of the applied non-equilibrium stage model for RD. The validated process model could be used for an intensive theoretical analysis of the reactive separation process. The non-equilibrium stage model is implemented in the simulation ASPEN Custom Modeller® environment, which considers reaction kinetics, multi-component mass and heat transfer rates as well as column hydrodynamics. As the homogeneously catalysed transesterification was carried out in a packed column, the reactive distillation model also takes into account the chemical reaction in the liquid hold-up of the internal column distributors. Consequently, the validated process model was used to determine the influence of operational parameters on conversion and selectivity.

(107e) Phenomena-Based Process Synthesis and Design to Achieve Process Intensification

Philip Lutze, Department of Chemical & Biochemical Engineering, Denmark Technical University, Lyngby, Denmark, Rafiqul Gani, Chemical and Biochemical Engineering, Technical University of Denmark, Lyngby, Denmark and John M. Woodley, Chemical and Biochemical Engineering, Technical University of Denmark, Kgs. Lyngby, Denmark

In recent years, Process Intensification (PI) has attracted considerable interest as a potential means of process improvements and to meet the increasing demands for sustainable production. PI aims to improve processes by increasing efficiency, reducing energy consumption, operational costs, volume, and waste as well as simplifying the flowsheet. A variety of intensified operations and equipment has been developed. Potentially, this creates a large number of options for possible process improvement, however, to date only a limited number of intensified technologies have achieved implementation, such as reactive distillation, dividing wall columns and reverse flow reactors [1]. One major reason for this is that the identification of the best PI option is neither simple nor systematic.

In previous work [2] we reported the development of a general computer-aided systematic synthesis and design methodology in which redundant intensified options are systematically removed by checking against predefined constraints through a decomposition approach of a superstructure optimization problem. In this approach lower level steps employ simple and easy calculations, while the higher level steps employ more rigorous and detailed calculations. However, up to now, this methodology is limited to already reported PI unit operations which can be retrieved, together with all information necessary for synthesis and design of each of them, from a knowledge base tool. In order to invent new unit operations going beyond those currently in existence, process synthesis and design incorporating PI needs to be investigated at the phenomenological level [3, 4] which will be presented in this contribution. The basis of the phenomena-based process synthesis is transfer units consisting of mass, component, energy and momentum balances as well as phenomena building blocks and model equations describing them. Following the same rationale as the general methodology described previously, the problem is first defined, given a set of product quality and quantity as well as additional process constraints in addition to the potential need for the improvement. Next, the given information is analyzed to identify all potential phenomena building blocks. In the next step, the phenomena building blocks are joined together according to combination rules to match process intensification targets defined through a superstructure of flowsheet options. Based on these, the generated options are screened through performance specifications before unit operations are identified. For example, a counter-current sequence of similar simultaneous mixing and vapour-liquid phenomena with the final steps at both ends being simultaneous heat transfer, mixing and vapour-liquid phase change phenomena can be identified as a distillation column. In the penultimate step, the remaining options are optimized with respect to a defined objective function. The results are verified through rigorous model simulations in the final step. The advantage of the phenomena-based process synthesis and design is that it generates potentially novel process options because the initial search space is wider than the search space of existing units (truly predictive models lead to reliable predictive solutions) as well as the simultaneous development of the necessary process models.

In this contribution the application of a phenomena-based process synthesis combined with a systematic methodology through a computer-aided framework will be described and highlighted with selected examples in the area of distillation systems, together with a focus on some of the suitable methods and tools.

[1] Harmsen, J. Process Intensification in the petrochemicals industry: Drivers and hurdles for commercial implementation. Chem Eng Process 2010; 49, 1, 70-73.

[2] Lutze, P., Gani, R., Woodley, J.M. Process Intensification: A perspective on process synthesis. Chem Eng Process 2010; 49, 6, 547-558.

[3] Freund, H., Sundmacher, K. Towards a methodology for the systematic analysis and design of efficient chemical processes: Part 1. From unit operations to elementary process functions. Chem Eng Process 2008; 47, 12, 2051-2060.

[4] Papalexandri, K.P., Pistikopoulos, E.N. Generalized modular representation framework for processs synthesis. AIChE J 1996; 42, 4, 1010-1032.

See more of this Session: Dr. Fair's "Enhanced Distillation": Advances in Thermally Coupled and Reactive Distillation
See more of this Group/Topical: Topical 8: The Dr. James Fair Heritage Distillation Symposium

(107f) Control Structure Design for Optimal Operation of Thermally Coupled Columns

Deeptanshu Dwivedi, Deparment of Chemical Engineering, Norwegian University of Science and Technology, Trondheim, Norway, Ivar J. Halvorsen, ICT, SINTEF, Trondheim, Norway and Sigurd Skogestad, Department of Chemical Engineering, Norwegian University of Science and Technology, Trondheim, Norway

Control Structure design for optimal operation of thermally coupled columns

Deeptanshu Dwivedi[1], Ivar J. Halvorsen[2] and Sigurd Skogestad[1]

[1] Department of Chemical Engineering, Norwegian University of Science and Technology, N-7491 Trondheim, Norway, Email: dwivedi@nt.ntnu.no, skoge@nt.ntnu.no

[2] SINTEF ICT, Applied Cybernetics, N-7465 Trondheim, Norway, Email: Ivar.J.Halvorsen@sintef

Keywords: Thermally coupled Distillation Columns, Optimal Operation, Control Structure Design, Divided Wall Column

Abstract:

The dividing wall column (DWC) has gained increased industrial attention due to its energy and capital-saving properties. To obtain the energy-savings in practice, it is required to apply a control strategy that keeps the operation close to the optimal operation in presence of unknown disturbances and model uncertainties. Otherwise, the potential energy savings may easily be lost or lower purity may result, especially in the side stream product. The three-product DWC has been used in more than hundred industrial applications. As we increase the complexity by going to the 4-product Kaibel column and even the 4-product DWC with two partition walls, and use the column to separate feeds with a large number of components which shall be grouped into suitable product streams, the question of control design and control performance will become even more important.

Minimum energy operation and tight control are closely related. In a conventional two product column, the most common control strategy is a simple one-point temperature control. With a large the energy input, in event of usual disturbances during operation, this very simple strategy can ensure purity in both products. This however, usually results in either both or one of the products is over purified. To obtain minimum energy operation, it is well known that both products should be controlled to their purity specifications. This is actually a more difficult mode of operation.

This work aims to demonstrate the control structure design and performance by operation of a four-product lab scale Kaibel column using the experimental runs followed by modelling and identification. A separation of mixture of four alcohols: methanol, ethanol, propanol and butanol are studied using the lab column. The lab column is a made of standard vacuum glass sections with 50 mm internal diameter and filled with 6 mm Rachig rings. The height of the set-up is about 8 m. There are 24 temperature sensors inside the column and the column is operated using a Labview interface. The column has a magnetic funnel that divides the liquid between the pre-fractionator and the main column. The vapour split is affected using two motor operated butterfly valves, one each in the pre-fractionator and in the main column. The reboiler is a 2 KW kettle type electrical heater.

A control structure using four decentralised PI temperature controllers is proposed. One temperature in the pre-fractionator and three temperatures in the main column were controlled using four manipulated variables namely liquid split ratio and 3 product flow rates: distillate, side product 1 and side product 2. The column was successfully stabilized and all the four temperature were controlled within ±0.5 C of the set-point. The product samples were collected and analyzed. The preliminary experiments yielded significantly pure top and bottom product while the two side products were not very pure. A suitable temperature profile should be maintained using the 4 PI decentralized loops to improve the purities. This is being further studied using a rigorous model of the same and lab experiments.

An effective vapour split is critical during minimum energy operation of the column. The vapour split valves were examined and an experiment was designed to test their efficacy. The column was run under total reflux conditions using only two components: methanol and ethanol. The vapour split valves are operated under split range control to set a constant temperature difference between a sensor in the pre-fractionator and one in the main column. This control objective was successful tested for servo and regulatory performance.

The model of the lab setup is a rigorous first principle equilibrium stage based model. There is no vapour hold up and model accounts for non-ideality in liquid phase using Wilson equation. The number ideal stages in the real column is determined experimentally by doing a total reflux experiment and using only two components, methanol & and ethanol. The compositions of the product sample were then analyzed to determine the number of ideal stages using Fenske Equation. The next step in the identification was to manipulate the inputs in the model to match the experimental conditions and data reconciliation. Here we fit the steady state temperature profiles, and the compositions of the product samples collected during the steady conditions of operations of the lab column. An optimization problem was formulated for the same [2]. A good fit of temperature snap shot from the steady operation condition of the experiment was obtained. The dynamic response of the experiments can be further fit using other parameters of the model like stage hold up and sizing.

This model is then used to test the efficacy of control structure for its regulatory response. The Kaibel column, owing to its highly interactive nature has some unique control problems and demand novel control solutions [3]. The key is to stabilize each internal sub-column profile to ensure that the separation task in each sub-column is carried out properly in an event of disturbances like feed property variations, tray/packing performance variations, measurement uncertainties and uncertainties in setting the internal flow splits. This can be done by use of feedback control, based on knowledge of how the internal sub-columns should behave at and around the minimum energy operating region.

Here we will show a procedure based on characterizing the optimal operating region. The V-min diagram [4, 5] is a simple tool that can be used to assess both design and operation issues. It can be point out the overall minimum energy requirement and flow distribution in fully thermally coupled arrangements. Further, we also get information on allowable variation margins in operation of the internal sections or sub-columns of the arrangement without compromising the overall minimum energy operation. This is very important for practical implementation. Due to inevitable design uncertainties and unknown disturbances an industrial column control system has to be designed to allow for certain slacks and the information about how and where operation can be relaxed is important in order to design the control strategy. For instance, an allowable range for vapour splits, allowable variation in feed properties for a selected control strategy, or if the focus should be on controlling the upper or lower part of the pre-fractionator profile. By identifying the key operation parameters that need high attention and thereby accurate control, we obtain a system that can give high purity in all products with a simple feedback control strategy.

References:

1. Kaibel, G. (1987). Distillation Columns with vertical partitions, *Chem. Eng, Tech. 10 (1987) 92-98*

2. Lid, T. Skogestad, S (2008). Scaled steady state models for effective on-line applications, *Comp. & Chem. Engg.Vol. 32, Issues 4-5, April 2008, Pages 990-999*

3. Strandberg, J., Skogestad, S., Halvorsen I. (2010), Practical control of dividing wall column, *Distillation & Absorption (Eindhoven, Netherlands)*

4. Halvorsen, I., Skogestad, S. (2006), Minimum Energy for the four-product Kaibel-column, *AICHE Annual meeting, 216d*

5. Halvorsen, I., Skogestad, S. (1999), Optimal Operation of Petlyuk Distillation: Steady State behaviour, *Journal of Proces. Control, vol 9, 1999, 407-424*

See more of this Session: Dr. Fair's "Enhanced Distillation": Advances in Thermally Coupled and Reactive Distillation
See more of this Group/Topical: Topical 8: The Dr. James Fair Heritage Distillation Symposium

Paper 107g

HEAT INTEGRATED REACTIVE DISTILLATION

C. B. Panchal **, Jing Huang, Robert W. Lyczkowski, and Richard D. Doctor
Argonne National Laboratory
Argonne, IL 60439
CPANCHAL@anl.gov

John C. Prindle
Tulane University
New Orleans, LA

Emmanuel Dada
ChemProcess Technologies, LLC
League City, TX

Philip Lutze, Rafiqul Gani, John M. Woodley
Denmark Technical University
Lyngby, Denmark

** To whom correspondence should be sent

Key Words: reactive distillation, heat integration, thermal management, energy efficiency

Prepared for Presentation at the 2011 National Meeting, Chicago, IL March 14-17, 2011

Unpublished

HEAT INTEGRATED REACTIVE DISTILLATION

C. B. Panchal[1], John C. Prindle[2], Jing Huang[1], Robert W. Lyczkowski[1], Richard D. Doctor[1], Emmanuel Dada[3], Philip Lutze[4], Rafiqul Gani[4], and John M. Woodley[4]

1. Argonne National Laboratory, 9700 S. Cass Avenue, Argonne, IL 60439 USA
2. Lindy Boggs Center, Dept. of Chemical and Biomolecular Engineering, Tulane University, New Orleans, LA 70118 USA
3. ChemProcess Technologies, LLC, 1916 Lake Landing Drive, League City, TX 77573 USA
4. Technical University of Denmark, Dept. of Chemical and Biochemical Engineering, Søltofts Plads, DK-2800 Kgs. Lyngby DENMARK

Abstract

Significant advances have been made by industry and academic research organizations to establish pathway to industry-wide applications of Reactive Distillation (RD). It is an effective but challenging process for energy-efficient synthesis of major commodity as well as high-value chemicals in emerging markets. However, with the phase-out of ethers as additives in gasoline the number of actual reactive distillation installations is quite limited despite a large number of patents and extensive academic and industrial research. Primary reasons that are perceived are: a) the difficulty of matching the optimum conditions for combined reaction and separations, that is, the optimal heat and pressure for separation may not match the optimal heat and pressure for carrying out the reaction; b) uncertainties in validated design models for the scale-up of laboratory data to commercial units; c) effective catalyst loading to maximize the effectiveness and d) thermal management of reactions with large exothermic and endothermic reactions. The project focuses on these crucial design issues with a goal of expanding industrial application of reactive distillation.

Introduction

Energy used to drive separation processes accounts for approximately sixty percent of the total energy used by the chemical and petroleum industries. The amount of energy consumed by distillation processes is much larger than the energy used to drive all other separation processes combined [Ozokwelu, 2005]. Conventional distillation remains the dominant separation process for the chemical industry even though it is less than 10% efficient and accounts for a high energy consumption exceeding 2,000-trillion BTU/yr in the U.S. Significant work has been done by industry and universities to establish the commercial advantages of Reactive Distillation [Hiwale, 2004]. Reactive Distillation in its simplest form combines exothermic chemical reaction heat to drive product separation using distillation.

In theory, reactive distillation can be applied to wide range industrial processes described in a comprehensive study of Sharma and Mahajani [2003]. Further, Harmsen [2007] pointed out that reactive distillation is a recognized unit operation and can be considered as a front-runner in process intensification. According to this review study there are about 150 commercial installations of reactive distillation worldwide; nearly one half of which for production of ethers (MTBE, TAME and ETBE). Harmsen, in a consistent manner with other investigators, listed governing techno-economic factors key to industry wide applications of reactive distillation. Reactive Distillation has established itself as a technique that can significantly improve chemical industry performance and competitiveness by reducing capital costs, process energy use, and environmental impacts. [Sundmacher, 2003].

The focus of this project consists of: a) thermal management for optimum global temperature profiles that would influence distribution of reactants and products and local temperature distribution in the catalyst bed; b) effective ways to improve contacting of the reactive species with the catalyst; and c) a rate-based model to account for localized reactive multi-phase flows initially, the thermodynamics will consider equilibrium reactions but they later will be extended to non-equilibrium conditions. Specifically, the experimental effort for this project will focus on improved contacting and wetting of the catalyst on the reactive distillation trays and meeting the challenge of developing a heat transfer device that will not adversely impact the distillation hydraulics. [*see figure 1 below*].

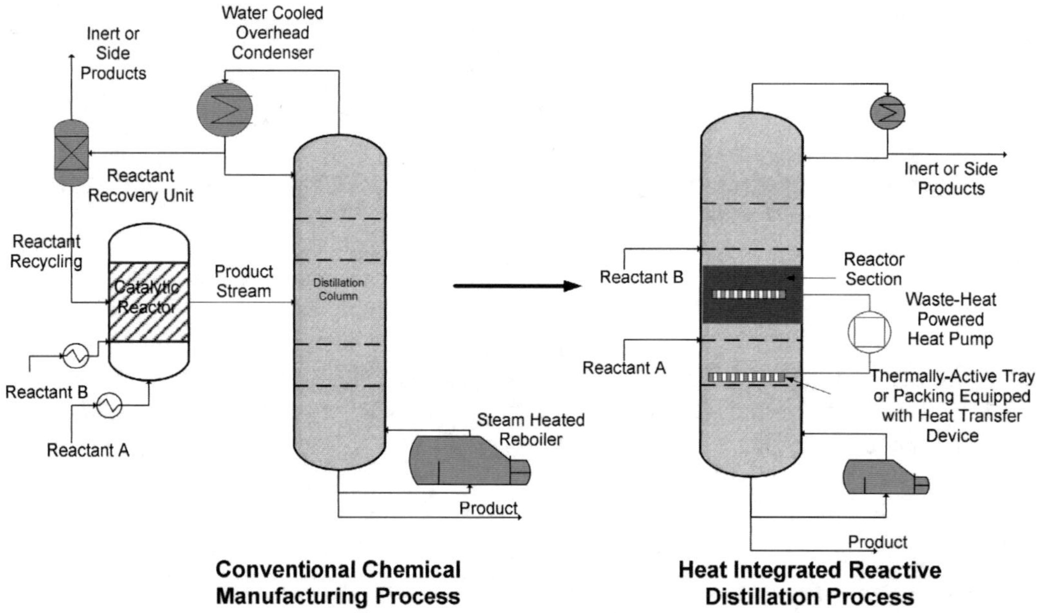

Fig. 1 A comparison of conventional manufacturing with Heat-integrated reactive distillation

Technology Readiness Level (TRL) Assessment

In order to realize the full potentials of reactive distillation, it is essential to develop methodology to determine applicability of reactive distillation to individual or class of chemical synthesis. It should be borne in mind that techno-economic feasibility proven for one industrial process cannot be generalized to other processes due to the complexity of reactive distillation. This can be achieved by performing *Technology Readiness Level* (TRL) analysis that is commonly applied to emerging technologies. In this project we are pursuing TRL analysis with the focus on the following factors: a) energy efficiency and water footprint; b) catalyst loading and innovative ways to support/package catalysts; c) thermal management; d) scale-up (rate-based model supported by pilot-plant data); e) operation (start-up, shutdown, and normal) & control strategy; and f) CAPEX – capital expenses (new plant vs. retrofit).

The TRL analysis is being applied to manufacturing of large-scale commodity chemicals as well as high-value chemicals for emerging markets. Based on the TRL-based merit-criteria analysis this project developed a priority list that is the focus of this project. The priority list consists of: a)

hydration/dehydration processes (e.g., cyclohexanol and ethylene glycol); b) alkyl carbonates, electrolytes for battery storage applications (e.g. dimethyl carbonate); and c) isomerization (e.g., n-C_6-C_8 to iso-C_6-C_8).

Integrated Approach

An integrated approach of this project consists of: a) process analysis to evaluate energy and economics; b) CFD analysis to map-out liquid and vapor flows in Katapak-SP11 structured packing; c) pilot-scale tests for validating the system model; d) integration of thermal management and the opportunities to add or remove heat for energy efficiency and process optimization; and e) control strategy using the heat-integrated distillation column (HIDiC) design. This project builds upon the developments reported in this field and we have compiled an extensive literature base on the reactive distillation. We intend to fully utilize and build-upon the past work in an effective manner to advance the reactive distillation technology by moving up TRL of the focus industrial processes mentioned above. If successfully demonstrated, this would serve as a prototype for other industrial processes of potential applications of reactive distillation. In order to achieve this goal, we have assembled a unique Team with multi-disciplinary capabilities.

The project focuses on heat-integration to expand the applicability of reactive distillation by providing active methods to achieve and maintain optimum conditions for separations, reaction, and scale-up of pilot-scale data to design of commercial units. The investigation consists of: a) thermal management for optimum global temperature profiles that would influence distribution of reactants as well as localized temperatures, including within catalyst packages; b) effective ways to improve contacting of the reactive phases with the catalyst; c) a rate-based model to account for localized reactive multi-phase flows; and d) pilot-scale tests.

Thermal Management

The basic objectives of reactive distillation thermal management are: a) to improve energy efficiency; b) to improve conversion and yield while reducing undesirable side products; c) to reduce catalyst loading for a given conversion; and d) to reduce cooling water demands. In particular, the opportunities for significant energy efficiency improvements with Heat-Integrated Distillation Columns (HIDiC) are being actively pursued and can be achieved as demonstrated by Ohe [2007].

Heat integration in distillation processes is a common practice. The primary focus is on the following technical approaches:

a. Feed-effluent heat exchanger network: as for example in crude-oil preheat trains;
b. Pump-around systems: as for example, top and bottom pump-arounds in the primary crude still;
c. Cascaded distillation columns: as for example in ethanol separation;
d. Mechanical vapor recompression (MVR); for example, C_3/C_4 fractionation;
e. Thermally-active trays and packed columns: limited to theoretical analysis;
f. Heat-integrated columns: rectification and stripping sections operating at different pressures; and
g. Waste-heat powered thermal heat pump which is another emerging technology.

Investigators have been pursuing heat integration in reactive distillation. Michigan State University and the Denmark Technical University as part of this program effort have prepared a list of recent publications on heat integration.

A summary of findings from the literature is as follows.

a. Equilibrium-controlled reactions are generally considered for heat integration.
b. Heat integration is primarily considered for highly exothermic reactions to remove the heat of reaction that may adversely impact the V/L equilibrium and hence adversely impact reactant/product concentration profiles along the column, one such example is the hydrolysis of ethylene oxide to ethylene glycol.
c. Endothermic reactions in general require heat integration. Considering the complexity, no significant developments have been reported using heat-integrated reactive distillation for endothermic reactions with high heat of reaction.
d. For moderate heats of reaction, in general heat integration is not considered for improved energy efficiency as being done for conventional distillation processes.
e. No detailed analysis or experimental measurements of temperature distribution with catalyst bed/package and in the vicinity of interface with vapor flow path.
f. Split-column heat integration offers significant advantages for optimizing reactions, here reactions in the rectification and stripping sections will be operating at two different pressures and hence temperatures can be optimized independently.
g. A 10% to 20% reduction in reboiler and condenser heat duties is reported based on systems analysis of internal heat integration; however, no supporting test data have been publically disclosed.
h. Advantageous features of an active heat integration system on dynamic controls of reactive distillation have been recognized. However, once again no supporting test data have been reported.

The application of ***an active thermal management system*** in reactive distillation depends on the following merit criteria:

- The heat of reaction (exothermic or endothermic) relative to the heat of vaporization;
- Reactive distillation: heat/mass transfer controlled; kinetically controlled or equilibrium controlled;
- Dependency of reaction selectivity on local temperature;
- Transient responses of reactive distillation to process conditions: such as the reflux ratio and reboiler operation; and
- Economic incentives to apply heat integration to improve energy efficiency with or without reactive distillation.

In this project we are developing merit criteria to determining appropriate configurations for several classes of equilibrium reactions based upon the relative volatility of the reactants and products, heat of reaction, and the activation energy reflecting dependency of kinetics on local temperatures. A basic assumption for this analysis is that for exothermic reactions, the optimal conditions are at the lowest feasible temperatures (i.e. lowest feasible operating pressure). The corollary is true for endothermic reactions. Both of these assumptions are based upon thermodynamics for equilibrium reactions but they later will be extended to non-equilibrium conditions. When applying this assumption in determining whether external heat integration will increase efficiency, the following approach is taken. For exothermic equilibrium reactions, the trays higher in reactant concentrations should have higher temperatures than those for trays higher in product concentrations. This way, the reverse reaction rate is lowered in the high-product regions of the reaction zone. For endothermic equilibrium reactions, the corollary is true – namely that trays higher in reactant concentrations should have lower temperatures than those for trays higher in product concentrations. Again, this methodology will reduce the reverse reaction rate in the high-product regions of the reaction zone.

This project focuses on the following three technology developments:

1. Thermally-active trays and packed columns;
2. Heat integrated columns; and
3. Waste-heat powered thermal heat pumps.

Inter-stage heat transfer devices can be readily incorporated into packed columns, which could also help to re-distribute liquid and vapor phases [*see Figure 2a*]. Thermally-active trays could include heat-transfer devices on the tray, above the tray in the froth/mist region or in the down-comer, depending on reaction types [*see Figure 2b and 2c*]. New innovative down-comer system may have to be developed to accommodate heat transfer devices on the tray for thermally-active tray concept. Heat-transfer devices can be tube coils with appropriate enhancements or compact brazed/welded units. The development of heat-transfer device that would exchange heat effectively without adversely affecting the hydrodynamics is very important. Furthermore, the development of heat-transfer devices is an essential step for heat integrated columns, where heat would be transferred from one section of column to the other section. Waste-heat powered thermal heat pump can play key role in reactive distillation by improving energy efficiency and developing active controls.

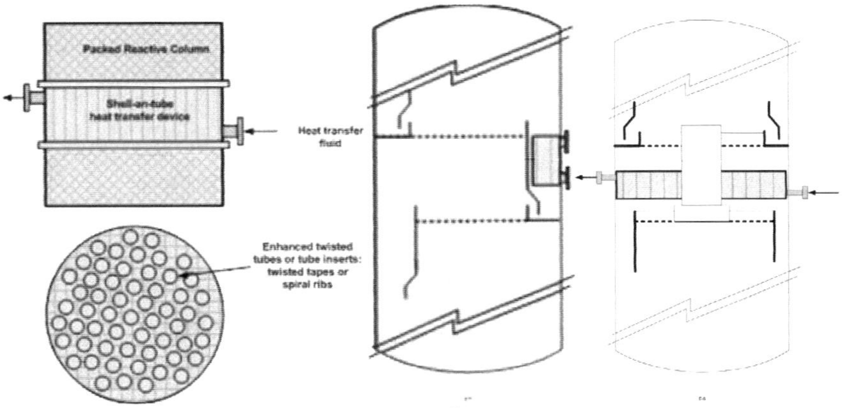

a b c

Figure 2. Conceptual heat transfer devices: a) inter-stage; b) down-comer; and c) froth/mist flow device.

Aspen+ analysis is being performed to optimize the internal heat management for the reaction of cyclohexene with formic acid, which is a possible first stage for the production of cyclohexanol. The second stage consists of the hydrolysis of cyclohexene formate. Figure 3a shows a simple example of transferring heat from one of the stripping section stages to one of the rectification stages, which does not require outside heat-pump system. The Aspen+ results illustrate the significance of thermal management in reactive distillation. In reality, heat would be transferred from rectification section to the stripping section, either using thermal heat pump or heat-integrated split column design with optimized locations of reactant feeds. The systems analysis will focus on three key design parameters: a) catalyst loading; b) conversion; and 3) reboiler duty, as illustrated in Figure 3b.

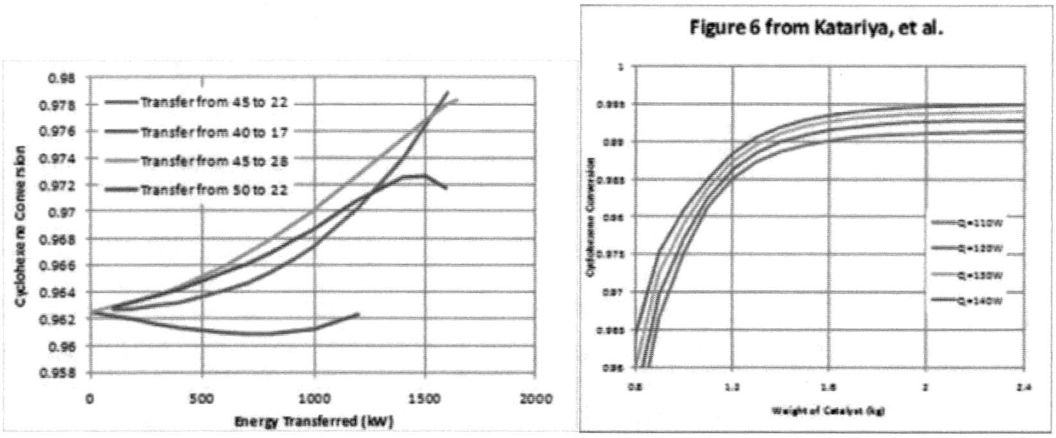

a b

Figure 3. Parametric analysis to determine optimum internal heat integration of cyclohexanol process

Conclusions

We have identified and characterized some of the critical barriers, arising from complexity of combining reactions and separations in one unit, that are critical to industry wide application of heat-integrated reactive distillation. We contend that "brute force" has been employed to design selected commercial scale units based on pilot plant data; however, validated "rate-based" models should be pursued in order to integrate the developments of individual subsystems, such as packing, trays, downcomer configurations, compact heat transfer devices, and better understandings of reactive multiphase flows. Pilot-scale experiments with detailed measurements along the length of the columns are an essential part of the development of the rate-based model. Dynamic controls of reactive distillation columns, including start-up and shutdown, have been recognized as critical areas for further developments. An active thermal management is expected to help in developing controls to maintain process parameters within optimum range of operation that include distributed internal reflux/reboiler which can minimize the impact of variations in overhead reflux caused by cooling media and performance of condenser.

Acknowledgements

This work supported by the U.S. department of Energy, Assistant Secretary of Energy Efficiency and Renewable Energy, and the Office of Energy Efficiency and Renewable Energy, Industrial Technology Program, under Contract DE-AC02-06CH11357 by UChicago Argonne, LLC with the U.S. Department of Energy. The authors acknowledge the support of Dr. Dickson Ozokwelu, Technology Lead, Chemical Industries.

REFERENCES

Harmsen GJ, 2007, "Reactive distillation: The front-runner of industrial process intensification: A full review of commercial applications, research, scale-up, design and operation," *Chemical Engineering and Processing,* 46, pp. 774-780.

Hiwale, Rameshwar S.; Bhate, Nitin V.; Mahajan, Yogesh S.; and Mahajani, Sanjay M. [2004] "Industrial Applications of Reactive Distillation: Recent Trends," *International Journal of Chemical Reactor Engineering*: Vol. 2: R1.

Ohe S, 2007, "Energy-Saving Distillation through internal heat exchange (HIDiC)", *AIChE Distillation 2007: Topical Conference Proceedings, 2007 Spring Meeting*, Houston, TX.

Ozokwelu, D. et al. [2005-April], Hybrid Separations/Distillation Technology: Research Opportunities for Energy and Emissions Reduction, DOE Industrial Technology Program.

Sharma MM and Mahajani SM, 2003, "Industrial application of reactive distillation," *Reactive Distillation: Status and Future Directions*, Edited by Kai Sundmacher and Achin Kienle, Publisher Wiley-VCH, Weinheim, Germany.

Sundmacher, K. and A. Kienle [2003], Reactive Distillation: Status and Future Directions, Wiley: VCH.

(107h) Composition Estimation In DWC Columns Using Temperature Measurements

Maryam Ghadrdan, Norwegian Univeristy of Science and Technology, Trondheim, Norway, Ivar J. Halvorsen, ICT, SINTEF, Trondheim, Norway and Sigurd Skogestad, Department of Chemical Engineering, Norwegian University of Science and Technology, Norwegian Univeristy of Science and Technology, Trondheim, Norway

Composition Estimation in DWC Columns Using Temperature Measurements

Maryam Ghadrdan[1], Ivar J. Halvorsen[2] and Sigurd Skogestad[1]

[1] Department of Chemical Engineering, Norwegian University of Science and Technology, N-7491 Trondheim, Norway, Email: ghadrdan@nt.ntnu.no, skoge@nt.ntnu.no

[2] SINTEF ICT, Applied Cybernetics, N-7465 Trondheim, Norway, Email: ivar.j.halvorsen@sintef.no

Abstract

In this work, we propose a method to estimate the product compositions in a four product dividing wall column based on a combination of a number of temperature measurements.

Keywords: Kaibel column, Thermally coupled column, composition estimation, combination of measurements

Introduction

The Kaibel distillation column is considered as a intensified process which can replace three columns and separate a feed to 4 products. This four product divided-wall distillation column (DWC) contains fully thermally coupled sections built into a single shell. This arrangement is so interesting for strongly reduced energy consumption and construction costs. The tight integration makes it challenging to control, compared to the conventional sequence of simple columns.

It is critical to have a good estimate of product compositions. Reliable and accurate measurement of product compositions is one of the important issues in distillation column control. On-line composition measurement devices are expensive and not very reliable to be used directly in closed loop control and there is usually a considerable time delay that may be a limitation to control performance. Temperature measurements are fast, inexpensive and more reliable and have been used for distillation column control in industry instead of composition analyzers. Mejdell and Skogestad [1] have mentioned different reasons of inefficiency of single temperature measurement in distillation columns in their study. Composition changes in feed, the effect of variation of off-key components, noise in measurement devices, temperature variation due to flow pulses and improper mixing on the trays, pressure changes are named as sources of inefficiency. Some of them can be compensated. The only problem which can not be corrected is to get constant composition by fixing a single temperature some trays away from it. They have suggested a method for estimating the product compositions by measuring temperatures of all trays.

In this work, we propose an alternative approach for designing estimator which is to use the self optimizing control strategy and find a combination of temperature measurements in the Kaibel column. This work is a continuation of the work done by Hori et al. [2]. In this work we will include noise. The number of measurements which result in one control variable depends on the number of temperature sensor locations which are put in the column during construction. By keeping the combination of temperatures constant, we can make sure that the process is optimal even after disturbances occure and therefore the compositions will remain constant as the optimal steady state values. This approach will be compared with the Partial Least Square (PLS) approach proposed previously ([3-4]).

The idea behind self-optimising control is to find a variable which characterise operation at the optimum, and the value of this variable at the optimum should be less sensitive to variations in disturbances than the optimal value of the remaining degrees of freedom. Thus if we close a feedback loop with this candidate variable controlled to a setpoint, we should expect that the operation will be kept closer to optimum when a disturbance occur.

Self-optimizing control is when we can achieve an acceptable loss L with constant setpoint values c, for the controlled variables (Skogestad 2000).

Process Description

As mentioned above, the divided-wall distillation column (DWC) contains fully thermally coupled sections built into a single shell. The DWC is capable of separating three or four products with a single reboiler and condenser. The Kaibel column, which is a 4-product DWC, is shown in Figure 1. The two lightest and the two heaviest products are supposed to be separated in the prefractionator and the products are separated further and drained in the main column.

The temperature at a stage in a distillation column is a good indication of its composition. Skogestad [5] presents some benefits of using temperature loops for controlling the composition:

1. Stabilizes the column composition profile along the column

2. Gives indirect level control: Reduces the need of level control

3. Gives indirect composition control: Strongly reduces disturbance sensitivity

4. Makes the remaining composition problem less interactive and thus makes it possible to have good two-point composition control

5. Makes the column behave more linearly

The model has six degrees of freedom: boilup rate (V), reflux (L), side stream flows (S1, S2), liquid split (Rl) and vapour split (Rv), from which four will be used to keep the product compositions constant. There will remain two manipulated variables which are used as optimization variables.

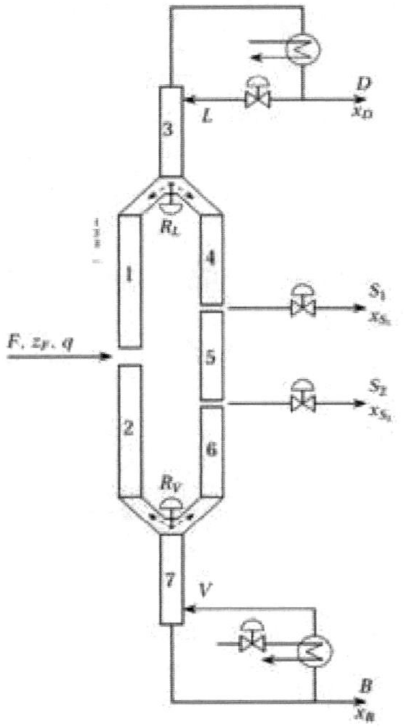

Figure 1. Schematic of a 4-product dividing wall column

The model used for this study is simulated in UNISIM. The feed stream is an equimolal mixture of Methanol, Ethanol, 1-Propanol, 1-butanol and saturated liquid. The optimal boilup is somewhat higher than the theoretical minimum boilup derived from minimum energy diagram proposed by Halvorsen et al. [6] which is with the assumption of infinite number of stages. This value is used as an initial estimate of the energy needed for a specified separation. All the optimal operating points for different sets of the disturbances are found by applying an optimisation solver in MATLAB with the full non-linear model in UNISIM.

References

1. Mejdell, T., Estimators for Product Composition in Distillation Columns. 1990, Norwegian University of Science and Technology.

2. Hori, E.S., S. Skogestad, and V. Alstad, Perfect Steady-State Indirect Control. Ind. Eng. Chem. Res., 2005. 44: p. 863-867.

3. Mejdell, T. and S. Skogestad, Composition Estimator in a Pilot-Plant Distillation Column Using Multiple Temperatures. Ind. Eng. Chem. Res., 1991. 30: p. 2555-2564.

4. Weber, R. and C. Brosilow, The use of secondary measurements to improve control. AIChE J. , 1972: p. 614-623.

5. Skogestad, S. and I. Postlethwaite, Multi-variable Feedback Control, Analysis and Design (2nd Edition). Wiley 2007.

6. Halvorsen, I.J. and S. Skogestad, Minimum Energy for the four-product Kaibel-column in AIChE Annual meeting 2006. 2006: San Francisco p. 216d

See more of this Session: Dr. Fair's "Enhanced Distillation": Advances in Thermally Coupled and Reactive Distillation

See more of this Group/Topical: Topical 8: The Dr. James Fair Heritage Distillation Symposium

(120a) Beer Tower Revamp Experience

Daniel R. Summers, Sulzer Chemtech USA, Inc., Tulsa, OK

Revamp experience in a beer distillation tower.

See more of this Session: Bio Fuels Distillation: Looking to the Future
See more of this Group/Topical: Topical 8: The Dr. James Fair Heritage Distillation Symposium

(120b) Alternative Separation Methods for Bioethanol Purification

Aristoklis Hadjitheodorou, School of Chemical and Metallurgical Engineering, University of the Witwatersrand, Johannesburg, South Africa and **Mark Peters**, Wits University, Johannesburg, South Africa

The environmental impacts of fossil fuel usage along with the depletion of available reserves have led scientists and engineers to seek greener and more renewable sources of energy. Bioethanol has proven to be one of the most promising alternatives to fossil fuel-derived petroleum as a transport fuel. Typical ethanol concentrations of fermentation from microorganisms are around 15%. This product is a very dilute solution therefore a large portion of the energy expenditure in the production of bioethanol is used to extract ethanol from the bioreactor effluent. Currently distillation is the most commonly used method of separation, however, it is ineffective in this situation as a large amount of energy is utilised in vapourising water which exits the column mainly as a bottoms product. Distillation is also limited as an azeotrope forms at 95%wt ethanol. Thus it is of significant interest to investigate other common separation methods in bioethanol purification, either as an alternative to distillation or as an addition, in an effort to reduce overall energy consumption. Column Profile Maps (CPM), originally developed for distillation systems, offers a graphical method for rapidly synthesizing complex separation configurations. The CPM technique, which is directly applicable to stripping and absorption processes, has been extended to other separation processes, such as membrane permeation and reactive distillation and can be developed for others (i.e. solvent extraction). Thus feasible hybrid systems can easily be synthesized from a similar basis and compared to the traditional method of distillation in terms of their 'ease of separation' and overall work efficiency.

See more of this Session: Bio Fuels Distillation: Looking to the Future
See more of this Group/Topical: Topical 8: The Dr. James Fair Heritage Distillation Symposium

Ethanol Distribution Over Fractions of Ethanol-Gasoline Fuel From Distillation and Spectroscopic Measurements

Sergey A. Karpov and Natalia B. Teterevyatnikova, Gubkin Russian State University of Oil and Gas, Moscow, Russia

The search for alternative sustainable fuels has grown enormously, especially in the 1990s as attention has been focused on environmental issues. At the forefront currently for additives to motor gasolines is ethanol. Ethanol is renewable and has the potential of reducing carbon dioxide emissions as compared to traditional fuels. It can be used in non-adapted, current vehicles at levels up to 10% by volume.

As it was mentioned above, environmental concerns have lead to the desire to add oxygenates to traditional gasolines. In the US and other parts of the world, methyl tert-butyl ether (MTBE) was accepted as the most economical oxygenate. Unfortunately, MTBE entered groundwater in some areas of the US, and its very low threshold for taste and odor have rendered the water undesirable for human consumption. The removal of MTBE opens a gap for an environmentally acceptable high-octane component for gasoline and is a key driving force in the search for alternative components. Ethanol is emerging as the additive to fill the MTBE gap and, it is being phased into California and the rest of USA where oxygenated fuels are required.

The addition of ethanol to hydrocarbons can significantly affect the volatility properties. Mixtures of ethanol with hydrocarbons exhibit large deviations from ideal mixing as defined by Raoult's law. Sometimes such non-ideal behavior can be reasonably predicted for binary systems using standard activity coefficient models and newly developed equations of state applicable for associating molecules (e.g. CPA and SAFT).

With the exception of octane number specifications, the most important properties defining current "boutique" gasolines are volatility properties. The key volatility properties are Reid vapor pressure and distillation.

It follows from the above-stated that it is necessary to understand the structure of alcohol containing gasoline components in order to resolve the problem of their excessive volatility.

It is obvious that when adding alcohol to gasoline it forms a number of azeotropes with gasoline hydrocarbons. Such azeotropes determine excessive volatility of alcohol containing gasoline components in comparison with the source gasoline. In this connection, it should be considered that different azeotropes between hydrocarbons were present in the system before adding of alcohol to the system.

Because of a complex fractional composition of gasoline we may not expect exact knowledge of composition and structure of all azeotropes, being formed by ethanol. At the same time, we may expect some general regularity. Particularly, a link should probably exist between group composition of the source gasoline and increase in vapor pressure when adding alcohol.

In order to find these general regularities, it is necessary to determine, how ethanol added is distributed between different fractions of gasoline.

The Pasquini paper describes the use of near infrared (NIR) spectroscopy for real-time monitoring of distillations. The high scan speed of an AOTF-based NIR spectrometer has been combined with a simple flow cell designed to collect the condensate whose absorbance spectrum is remotely acquired in real time during the distillation, through a pair of optical fiber bundles. The potential of the proposed system has been demonstrated via its use in the distillation of some laboratory-prepared binary mixtures of hydrocarbons, hydrocarbon and ethanol, and regular and adulterated gasoline samples.

Unfortunately NIR signal cannot be used for quantitative analysis of fuel fractions.

In Ref.[7] it is shown that, among the oxygenated compounds used as additives for gasolines, ethanol was the one that caused the most marked change in the distillation curve, through the formation of a threshold due to the minimum temperature azeotrope that was formed. The nonoxygenated compounds, isooctane and toluene, hardly altered the distillation curve.

Lack of quantitative model of ethanol distribution impedes all subsequent discussions at this stage. So this work is related to ethanol distribution over fractions of alcohol containing fuel.

Distillation of alcohol containing gasoline was performed in compliance with ASTM D 86 using manual distillation apparatus HDA 620 (HERZOR, Austria). This type of devices cannot be called the best for component separation, but their use is a standard in petroleum and related industries. Their results are used for estimation of quality of gasoline and other "light" petroleum products. So we have decided to use this type of device instead of any other (even more effective).

Distillation was performed at 5 °C intervals. Since – as our experience and Ref.[7] showed – ethanol significantly changes the shape of distillation curve, increasing its "sharpness" within 60-70 °C, the temperature range was varied from time to time with a view to collect a sufficient amount of fractions for subsequent analysis.

Determination of ethanol content in hydrocarbon mixture is possible using both gas chromatography method8 and infrared spectroscopy method. A good correlation between these two methods was demonstrated.0

As IR spectroscopy method is not so time consuming and doesn't require use of additional reagents, its use for analysis of 112 fractions, obtained in this work, is preferred. Moreover, application of attenuated total reflectance (ATR) add-on device makes the analysis even easier.

IR spectroscopy method was used for determination of ethanol content in gasoline and its fractions. Analysis was performed using Tensor 27 FT-IR Spectrometer (Bruker, Germany). MIRacle ATR add-on device (Bruker, Germany) with ZnSe crystal was used for analysis of fuel fractions. Detailed information on the unit and add-on device is given in Table 1 [3]. The obtained fractions were stored at the temperature of -18 °C not more than 24 hours since the moment of distillation.

Figure 2 [3] gives examples of distillation curve, obtained by us. As it is evident from this figure, relationships are analogous to those obtained before (see Ref.[1] and Fig. in Ref.[7]).

Addition of ethanol significantly changes distillation curves of gasoline. Ethanol leads to essential change in the curve shape – a plateau appears at about 68 °C. Furthermore, ethyl alcohol leads to essential increase in the quantity of low-boiling-point fractions. For example, if in the source gasoline (Gasoline A), whose distillation curve is given in Figure 2 [3] under 70 degrees Celsius 25% v/v of volume boiled off; after addition of 5% v/v ethanol under 70 degrees Celsius 51% v/v began to boil off. By increase in alcohol percentage to 10% v/v, this percentage increased to 56% v/v.

It was already found1 that the D-86 is similar to the TBP, having a "plateau" region during the ethanol–hydrocarbon azeotropes "region". The "step change" is now a "transition region" due to the sloppy separation of the D-86. The D-86 for our gasoline–ethanol blends has three distinct regions: plateau or azeotropic region, transition region, and dilution only region as already observed by authors of Ref.[1].

Figure 3 [3] illustrates influence of alcohol on distillation temperatures at percentage evaporated. One can see that addition of ethanol mostly influence on T10 and T50 points. The same situation is observed for Gasoline B. One should say that the change in alcohol concentration results in significant change of mostly T10 and T50.

Main objective of our work was to determine how alcohol is distributed over gasoline fractions. From Figure 2 [3] it is evident that the main part of alcohol isn't a clean gasoline component (characterized by own boiling point11 – 78. °C at 101. kPa). If alcohol didn't form azeotropes with gasoline hydrocarbons, its addition would result in a simple increase in quantity of fraction, boiling off within 75-80 °C. But the main influence of ethanol addition is below than that temperature limit.

This is not observed in Figure 2 [3]. Figure 2 [3] shows that alcohol forms a number of azeotropes with gasoline components, where most of these azeotropes has obviously a lower boiling point than the source hydrocarbons.

Examples of IR spectra (500-4000 cm-1) of two different fractions (with mean boiling temperatures of 47. and 77. °C) of alcohol containing fuel (Gasoline B + 10% v/v ethanol) are given in Figure 4 [3]. Figure 4 [3] shows that the maximum signal within the range of 3200-3600 cm-1 is caused by «light» gasoline fractions. Fractions with average boiling point over 90 °C have only traces of alcohol. These amounts are under our detection error for ethanol (0.% v/v), therefore we accepted the alcohol content of these fractions as zero.

The same fact was observed in Ref.[6] (see Fig. 6) by NIR method but no quantitative analysis was provided there.

Figure 5 [3] gives distribution of alcohol over different fractions of alcohol–gasoline composition. The maximal amount of ethanol is found in fraction 65-70 °C – up to 30% v/v. Low ethanol concentration is in the lowest boiling part of gasoline (IB-42 °C) and 85-90 °C – less then 5% v/v.

Attention is drawn to the fact of low alcohol content in its «native» fraction (with = 77. °C). So we can conclude that the main volume of alcohol is found in azeotrope form. It determinates the high volatility of ethanol–gasoline mixtures.

We showed in this work how exactly ethanol is distributed over different fractions of alcohol–gasoline blend. To our knowledge, this result is the first of its kind. Only qualitative data have been published previously (see "Introduction").

We found out that ethanol is concentrated in light fractions and is almost completely absent in heavy fractions (with boiling point over 90 °C). Its traces, found in fractions with high boiling points, should be probably referred to indistinct component separation during distillation as per ASTM D 86 (without reflux). The distribution form is like Gaussian distribution with maximum at ~60 °C.

Main part of ethanol is found to form azeotropes – the reason of the fuel high volatility. Only small part of alcohol (~3% v/v) "stays" in its "natural" form.

In the following, the use of these data14 – as we hope – will contribute to more precise determination of nature of increased vapor pressure for alcohol–gasoline blends and to solution of this problem, which will make alcohol–gasoline blends more competitive as compared to traditional gasolines.

References

1. S. A. Karpov (2008), "Aliphatic alcohols as components of fuels for internal combustion engines"

Chemistry and Technology of Fuels and Oils 44, 249.
2. S. A. Karpov (2008), "Current aspects of fuel ethanol production in Russia and abroad" Chemistry and Technology of Fuels and Oils 44, 1.
3. R. M. Balabin, R. Z. Syunyaev, S. A. Karpov (2007), "Molar enthalpy of vaporization of ethanol-gasoline mixtures and their colloid state" Fuel 86, 323.
4. R. M. Balabin, R. Z. Safieva (2007), "Capabilities of near infrared spectroscopy for the determination of petroleum macromolecule content in aromatic solutions" J. Near Infrared Spec. 15, 343.
5. Z. Nan, Z.-C. Tan, and L. Sun (2004), "Investigation on Thermodynamic Properties of Ethanol + Gasoline Blended Fuel" Energy Fuels 18, 84.
6. N. Peschke, S. I. Sandler (1995) "Liquid-Liquid Equilibria of Fuel Oxygenate + Water + Hydrocarbon Mixtures. 1" J. Chem. Eng. Data 40, 315.
7. S. Hellinger, S. I. Sandler (1995) "Liquid-Liquid Equilibria of Fuel Oxygenate + Water + Hydrocarbon Mixtures. 2" J. Chem. Eng. Data 40, 321.
8. R. M. Balabin (2010), "Is quantum chemical treatment of biopolymers accurate? Intramolecular basis set superposition error" J. Chem. Phys. 132, 231101.
9. R. M. Balabin, R. Z. Safieva, E. I. Lomakina (2010), "Gasoline classification using near infrared (NIR) spectroscopy data: Comparison of multivariate techniques" Anal. Chim. Acta 671, 27.
10. R. M. Balabin (2010), "Intramolecular basis set superposition error as a measure of basis set incompleteness: Can one reach the basis set limit without extrapolation?" J. Chem. Phys. 132, 211103.
11. R. M. Balabin (2010), "The identification of the two missing conformers of gas-phase alanine: a jet-cooled Raman spectroscopy study" Phys. Chem. Chem. Phys. 12, 5980.
12. R. M. Balabin (2010), "Boryl substitution of acetaldehyde makes it an enol: Inconsistency between Gn/CBS and ab initio/DFT Data" J. Phys. Chem. A 114, 3698.
13. R. M. Balabin (2010), "Conformational equilibrium in glycine: Experimental jet-cooled Raman spectrum" J. Phys. Chem. Lett. 1, 20.
14. M. B. Gramajo de Doz, C. M. Bonatti and H. N. Sólimo (2003), "Liquid–liquid equilibria of ternary and quaternary systems with two hydrocarbons, an alcohol, and water at 303. K: Systems containing 2,2,4-trimethylpentane, toluene, methanol, and water, or 2,2,4-trimethylpentane, toluene, ethanol, and water" Fluid Phase Equilibria 205, 53.
15. M. B. Gramajo de Doz, C. M. Bonatti and H. N. Sólimo (2003) "(Liquid+liquid) equilibria of ternary and quaternary systems with two hydrocarbons, an alcohol, and water at T=303. K. Systems containing cyclohexane, benzene, methanol, and water" The Journal of Chemical Thermodynamics 35, 825.
16. Jasem A. Alkandary, Adel S. Aljimaz, Mohamed S. Fandary and Mohamed A. Fahim (2001) "Liquid–liquid equilibria of water + MTBE + reformate " Fluid Phase Equilibria 187-188, 131.
17. R. M. Balabin (2009), "Tautomeric equilibrium and hydrogen shifts in tetrazole and triazoles: Focal-point analysis and ab initio limit" J. Chem. Phys. 131, 154307.
18. R. M. Balabin (2009), "Conformational equilibrium in glycine: Focal-point analysis and ab initio limit" Chem. Phys. Lett. 479, 195.
19. R. Z. Syunyaev, R. M. Balabin, I. S. Akhatov, J. O. Safieva (2009), "Adsorption of petroleum asphaltenes onto reservoir rock sands studied by near-infrared (NIR) spectroscopy" Energy Fuels 23, 1230.
20. R. M. Balabin, E. I. Lomakina (2009), "Neural network approach to quantum-chemistry data: Accurate prediction of density functional theory energies" J. Chem. Phys. 131, 074104.
21. R. M. Balabin (2009), "Polar (acyclic) isomer of formic acid dimer: Gas-phase Raman spectroscopy study and thermodynamic parameters" J. Phys. Chem. A 113, 4910.
22. R. M. Balabin (2009), "Enthalpy difference between conformations of normal alkanes: Raman spectroscopy study of n-pentane and n-butane" J. Phys. Chem. A 113, 1012.
23. J. Chen, L.-P. Duan, J.-G. Mi, W.-Y. Fei and Z.-C. Li (2000) "Liquid–liquid equilibria of multi-component systems including n-hexane, n-octane, benzene, toluene, xylene and sulfolane at 298. K and atmospheric pressure" Fluid Phase Equilibria 173, 109.
24. F. Aiouache and S. Goto "Liquid–liquid equilibria of ternary 2M1B/2M2B–2M1BOH–H2O and quaternary 2M1B–2M2B–2M1BOH–H2O mixtures" Fluid Phase Equilibria187-188, 415.
25. R. M. Balabin (2008), "Enthalpy difference between conformations of normal alkanes: Intramolecular basis set superposition error (BSSE) in the case of n-butane and n-hexane" J. Chem. Phys. 129, 164101.

26. R. M. Balabin (2008), "Intermolecular dispersion interactions of normal alkanes with rare gas atoms: van der Waals complexes of n-pentane with helium, neon, and argon" Chem. Phys. 352, 267.
27. R. M. Balabin, R. Z. Safieva, E. I. Lomakina (2008), "Wavelet neural network (WNN) approach for calibration model building based on gasoline near infrared (NIR) spectra" Chemometr. Intell. Lab. Syst. 93, 58.
28. R. M. Balabin, R. Z. Safieva (2008), "Motor oil classification by base stock and viscosity based on near infrared (NIR) spectroscopy data" Fuel 87, 2745.
29. R. M. Balabin, R. Z. Syunyaev (2008), "Petroleum resins adsorption onto quartz sand: Near infrared (NIR) spectroscopy study" J. Colloid Interf. Sci. 318, 167.
30. R. M. Balabin, R. Z. Safieva (2008), "Gasoline classification by source and type based on near infrared (NIR) spectroscopy data" Fuel 87, 1096.
31. I. Hatzioannidis, E. C. Voutsas, E. Lois, and D. P. Tassios (1998) "Measurement and Prediction of Reid Vapor Pressure of Gasoline in the Presence of Additives" J. Chem. Eng. Data 43, 386.
32. R. A. Harley and S. C. Coulter-Burke (2000), "Relating Liquid Fuel and Headspace Vapor Composition for California Reformulated Gasoline Samples Containing Ethanol" Environ. Sci. Technol. 34, 4088.
33. R. M. Balabin, "The First Step in Glycine Solvation: The Glycine-Water Complex" J. Phys. Chem. B 114, 15075 (2010).
34. R. M. Balabin, R. Z. Syunyaev, T. Schmid, J. Stadler, E. I. Lomakina, R. Zenobi (2011), "Asphaltene adsorption onto an iron surface: Combined near infrared (NIR), Raman, and AFM study of the kinetics, thermodynamics, and layer structure" Energy Fuels 25, 189.
35. R. M. Balabin, E. I. Lomakina, R. Z. Safieva, (2011) "Neural network (ANN) approach to biodiesel analysis: Analysis of biodiesel density, kinematic viscosity, methanol and water contents using near infrared (NIR) spectroscopy" Fuel, in press; doi: 10.1016/j.fuel.2010.11.038.
36. E. S. Mortikov, L. Kh. Kunashev, S.A. Karpov, M.A. Brykin (2009), "Preparation of high-octane components of gasoline from methanol over zeolite catalysts" Theoretical Foundations of Chemical Engineering 43, 779.
37. R. M. Balabin, R. Z. Syunyaev, S. A. Karpov (2007), "Quantitative measurement of ethanol distribution over fractions of ethanol-gasoline fuel" Energy Fuels 21, 2460.
38. R. M. Balabin, R. Z. Safieva, E. I. Lomakina (2007), "Comparison of linear and nonlinear calibration models based on near infrared (NIR) spectroscopy data for gasoline properties prediction" Chemometr. Intell. Lab. Syst. 88, 183.
39. A. V. Tsarev, S.A. Karpov (2009), "Increase of the environmental and operational characteristics of automobile gasolines with the introduction of oxygenates" Theoretical Foundations of Chemical Engineering 43, 563.

(120d) Reducing the Energy Demand of Cellulose Based Fuel Ethanol Through Salt Extractive Distillation Enabled by Electrodialysis

Mohammed A. Hussain and Peter H. Pfromm, Chemical Engineering, Kansas State University, Manhattan, KS

The Renewable Fuels Standard (RFS2), established under the Energy Independence and Security Act (EISA) of 2007, mandates the production of 36 billion gallons/year of renewable fuels in the U.S. in 2022, out of which 16 billion gallons/year must be produced from cellulosic sources. The fermentation based corn-to-fuel ethanol process is widely used to produce fuel ethanol in the U.S. While fermentation can be used to produce fuel ethanol from cellulosic sources, the concentration of ethanol in the fermentation broth may be less than 5 wt% compared to 10-15 wt% in case of corn-ethanol. Similar to corn-ethanol production, distillation can be used for separating and concentrating ethanol from fermentation broth followed by molecular sieve adsorption to achieve fuel grade purity, but the separation energy demand increases due to the drastic increase in distillation energy demand for ethanol concentrations less than 5wt%. 25,100 Btu/gal for 5 wt% ethanol compared to 15,100 Btu/gal for 15 wt % ethanol has been reported.[1] Dilute ethanol from the fermentation broth can be separated and concentrated aided by salt extractive distillation to directly produce fuel ethanol leading to significant energy savings. Techniques other than evaporation and drying for recovering salt, which is used to facilitate distillation, have rarely been considered. Since evaporative salt concentration/crystallization and solids drying techniques are energy intensive, reducing the energy demand for the salt recovery step becomes essential for reducing the overall energy demand. In this study, a combination of electrodialysis and spray drying is investigated for recovering salt. The salt extractive column bottoms stream is pre-concentrated by electrodialysis followed by spray drying. In electrodialysis, the dilute salt solution is concentrated by selectively removing the salt ions from the solution rather than evaporating water; therefore, requiring less energy than an evaporative process. Final recovery of dry salt is achieved in a spray dryer. Salt extractive distillation, with salt recovery enabled by a new scheme of electrodialysis and spray drying, is here conceptually integrated in the water-ethanol separation train of a fermentation based cellulosic fuel ethanol plant and investigated through process simulation using Aspen Plus® V7.2. Concepts, modeling, and results showing potential energy savings will be discussed.

[1]- Côté P, Noël G, Moore S. The Chatham demonstration: From design to operation of a 20 m^3/d membrane-based ethanol dewatering system. *Desalination.* 2010;250:1060-1066.

See more of this Session: Bio Fuels Distillation: Looking to the Future
See more of this Group/Topical: Topical 8: The Dr. James Fair Heritage Distillation Symposium

SIMULATION OF ETBE SYNTHESIS FROM TERT-BUTANOL AND ETHANOL BY BATCH RREACTIVE DISTILLATION USING HOMOGENEOUS CATALYSTS

Nada Bahjat Nakkash, Nadeen Khalid Al-Janabi
Al-Nahrain University, Baghdad – Iraq

Abstract

The present work concerned with studying the behavior of batch reactive packed distillation column to produce Ethyl tert-butyl ether (ETBE) by the reaction of tert-butanol (TBA) and Ethanol using two homogeneous catalysts H_2SO_4 and $KHSO_4$, experimentally and theoretically.

Experimentally, after checking the performance of catalysts and their ability to produce ETBE by bench experiment the reaction was carried out in a batch reactive packed distillation unit consists of heat resistance glass column of 71 cm total height and 3.5 cm inside diameter packed with glass rasching ring of 3 mm inside diameter at one atmosphere pressure.

The effect of many operating conditions on ETBE purity such as feed molar ratio of ethanol to tert-butanol (1:1, 2:1, 4:1), reflux ratios (3, 4, and 5), heat duties of (65, 90, and 146W), type of catalyst H_2SO_4 and $KHSO_4$, and the amount of catalyst have been studied, when the catalyst and reactants are mixed in the still. The best operating conditions when the reaction takes place in the still are FMR 1:1, reflux ratio 5, heat duty 146W and H_2SO_4 catalyst with 61.317% purity of ETBE and 94.789% conversion which are higher than using $KHSO_4$ catalyst even when three times its amount was used. For the best operating conditions an experiment was carried out using an intermediate reactive section with $KHSO_4$ catalyst placed in the middle section of distillation column. The results give a purity of ETBE 79.407% and 99.465% conversion which are higher than that obtained from reactive still but it needs large amounts of catalyst.

Theoretically, an equilibrium model (EQ) was developed using MATLAB (R2009b) to solve **MESHR** equations. The developed model results were compared with the results of an existed program ASPEN PLUS (10.2) for certain operating conditions, the deviation between the two programs is about $\pm$ 10% error. Results of experimental and theoretical model have been compared and show good agreement.

Keywords: ETBE synthesis, Simulation, Batch Reactive Distillation, Packed distillation.

Introduction

Reactive distillation (RD) is an alternative to a separated reaction-distillation processes, whereby reaction and separation take place with in a single counter current column. Reactants are converted to products in a reaction zone with simultaneous separation of the products and recycle of unused reactants to the reaction zone [1]. Among suitable reactive distillation processes is etherification process to produce octane enhancer such as production of ethyl tert-butyl ether (ETBE), methyl tert-butyl ether (MTBE), tert-amyl methyl ether (TAME), tert-amyl ethyl ether (TAEE) which are the best known oxygenates to be used to increase the octane level and to promote cleaner burning of gasoline and thus decrease harmful emissions from vehicles [2]. There is increasing interest in the use of ETBE as a replacement of methyl tert-butyl ether (MTBE) because of the latter's environmental problems [3], also ETBE has lower blend Reid vapor pressure (4 psi) than MTBE (9 psi) which allows ETBE to be used successfully in obtaining gasoline with less blend Reid vapor pressure than (7.8 psi) as required in some hot places during summer [4]. Furthermore, from environmental view point, ETBE is derived from ethanol which can be obtained from renewable resources like biomass [4].

Mostly ETBE has been synthesized by exothermic reversible reaction between isobutene (IB) and ethanol (ETOH), but the availability of isobutene is limited, it is only produced in refinery using catalytic and steam cracking operations, and it produces ETBE under high pressure. Therefore, alternative routes to synthesize ETBE are under substantial consideration. The most important substitute of isobutene is tert-butanol (TBA) which is a by-product of propylene oxide production in ARCO process.

In the synthesis of ETBE from TBA and ETOH, following sequence of reactions take place [5].

$$\underset{\text{TBA}}{(CH_3)_3COH} + \underset{\text{ETOH}}{CH_3CH_2OH} \longleftrightarrow \underset{\text{ETBE}}{(CH_3)_3COCH_2CH_3} + \underset{\text{Water}}{H_2O} \quad ...(1)$$

The above main reaction is accompanied by two side reactions

$$\underset{\text{TBA}}{(CH_3)_3COH} \longleftrightarrow \underset{\text{IB}}{(CH_3)_2C{=}CH_2} + \underset{\text{Water}}{H_2O} \quad ...(2)$$

$$\underset{\text{IB}}{(CH_3)_2C{=}CH_2} + \underset{\text{ETOH}}{CH_3CH_2OH} \longleftrightarrow \underset{\text{ETBE}}{(CH_3)_3COCH_2CH_3} \quad ...(3)$$

As isobutene exists only in the gaseous state at atmospheric pressure, so backward reaction in reaction (2) and reaction (3) can safely be neglected [4, 6].

Batch reactive distillation is suitable when the reaction products have lower boiling point temperature than the reactants [7]. For the present system (ETOH-TBA-ETBE-H_2O), ETBE has the lower boiling point therefore; in the present work batch reactive distillation process is considered.

Matouq et al. [8] synthesized ethyl tert-butyl ether from tert-butanol and ethanol in the presence of different acid catalysts ($KHSO_4$, $NaHSO_4$, H_2SO_4, and Amberlyst 15) at low alcohol grade (mixture of 80% mole percent water).

Sneesby et al. [9] presented a dynamic simulation model for ETBE synthesis from isobutene and ethanol. The simulation was developed using Speed Up simulation software, then the system was utilized for the study of transient open-loop responces and control system design.

Quitain et al. [10] synthesized ETBE from tert-butanol. Then, the effect of feed molar ratio of ETOH/TBA and type of catalyst on conversion and selectivity were studied. The experimental results were compared with results from ASPEN

PLUS simulator and it was shown good agreement.

Bisowarno and Tade [11] used dynamic simulation model to understand the effect of start up policy on the ETBE reactive distillation. This start up policy would results in the targeted ETBE purity within the shortest possible time without using excessive energy.

Yang et al. [12] developed a dynamic mathematical model for synthesis of ETBE from tert-butanol using ASPEN PLUS. The model incorporated reaction kinetics. The rapid solution for this model was obtained using Newton-Raphson method. Their simulation results showed good agreement with the literature results.

Jhon and Lee [13] presented dynamic simulation model using MATLAB software for reactive distillation column of 20 stages with ETBE synthesis. A structured and simple iterative algorithm was devised to estimate the vapor flows with the fast convergence under the rigorous energy balance. Both total and partial condensers are considered, and the reactive distillation performance for these condensers was observed. By comparison of their influence on purity of ETBE, total condenser is the better one.

Assabumrungrat et al. [4] used the experimental results of packed reactive distillation column for synthesis of ETBE from TBA to validate a simulation model using ASPEN PLUS software. They investigated the effect of various operating parameters such as condenser temperature; feed molar flow rate, reflux ratio, heat duty, and molar ratio of H_2O: ETOH on the reactive distillation performance.

Varisli and Dogu [14] produced tert-amyl ethyl ether (TAEE) and tert-amyl alcohol (TAA) by simultaneous etherification and hydration of 2-methyl-2-butene in a batch reactive distillation column. The effect of changing reboiler temperature in the range of (90-124) C° on total conversion was studied.

Hamza [15] studied a batch reactive distillation in packed column for esertification of methanol with acetic acid to produce methyl acetate using sulfuric acid as a homogeneous catalyst. The influence of various parameters such as batch time, reflux ratio, and feed molar ratio of methanol to acetic acid on performance of batch reactive distillation was studied, and the best conditions found are reflux ratio 2, feed molar ratio 2, and batch time 90mit.

Umar et al. [16] used a packed reactive distillation column to generate experimental data of ETBE synthesis from TBA. The effect of different key variables on reactant conversion and product purity in distillate was investigated. Umar et al. [16] studied the experimental synthesis of the same system with different macro porous and gelular ion exchange resin catalyst. Effect of various parameters such as temperature, reactants feed molar ratio, and catalyst loading were studied for optimization of reaction condition.

Nakkash et al. [17] presented a batch reactive multistage multi component distillation; rigorous model was used to build the simulation program using MATLAB to solve MESHR equations. The validity and accuracy of the developed program model were checked with previous work for methyl acetate production. The influence of various parameters, such as number of stages, batch time and liquid hold up on the performance of batch reactive distillation was shown through the vapor and liquid profiles, temperature, reaction rate, and composition profiles.

Present Work

The present work emphasized to construct mathematical model based on Equilibrium model (EQ) for packed batch reactive distillation. The proposed model predict the composition of each component, temperature, liquid and vapor flow and reaction rate for each step time of the batch time. The following assumptions were proposed for the developed model:

1. Vapor-liquid equilibria are achieved on each segment.
2. Constant pressure drop across the column.
3. Hold-up per segment equal to liquid hold-up on segment (i.e. vapor phase molar hold-up neglected).
4. Walls of column are perfectly insulated, therefore, the heat losses neglected.
5. Reactions take place in liquid phase.
6. Each segment is at mechanical and physical equilibrium.

The performance and validity of the developed model were compared with experimental results obtained throughout the present work (23) and with the results of an existed program ASPEN PLUS (10.2) for a certain operating conditions.

Many variables have been studied that affect the purity of ETBE and the % conversion such as feed molar ratio of ethanol to tert-butanol, reflux ratios, heat duties, type of catalyst and the amount of catalyst. All physical and chemical properties are taken from literature (23).

Experimental Work

Bench experiment to produce ETBE was carried out in a batch reactor, consists of 250ml two-necked round flask one handle thermometer and the other handling stir. Heat was supplied by a jacketed heating mantle with a constant heat duty of 220 W. The reactants, ethanol and tert-butanol has been mixed with a measured amount of catalyst, and the mixture was introduced to the flask, stirred, and heated to a temperature of 75 C°, two types of catalyst have been used ($KHSO_4$, and H_2SO_4). After a certain time a sample was taken and analyzed using Gas Chromatography with porpak Q column.

The results are given in Table (1) for $KHSO_4$ catalyst using feed volume ratio of ETOH: TBA (2:1) and after 3.5hr.

Table (1): Results of $KHSO_4$ Bench experiment

x_{ETOH}	0.738
X_{TBA}	0.152
x_{ETBE}	0.092
X_{H2O}	0.018

And for H_2SO_4 catalyst using the feed volume ratio of ethanol: tert-butanol (1:1) and after 4.5hr are shown in Table (2).

Table (2): Results of H_2SO_4 Bench experiment

x_{ETOH}	0.4232
X_{TBA}	0.0724
x_{ETBE}	0.2497
X_{H2O}	0.2360

A laboratory scale packed batch reactive distillation unit has been designed to perform the present work. The unit consists of a still pot consists of a three neck-round flask connected to the distillation column through its central opening. The other two openings were connected to a thermometer to measure the bottom temperature (vapor temperature), and a line for recycling a portion of solution in the distillate.

Heat for still pot was supplied by a jacketed heating mantle; its heat duty can be controlled using an electrical controller. Heat duties are 65, 90, and 146 W. The distillation column (located above the still pot) is made of a heat resistance glass column. To avoid the heat loss, the column was insulated with rubber insulation. The height of column is 71cm height and 3.5 cm inside diameter, packed with glass rasching ring of 10 mm length, 6 mm out side diameter, and 3 mm inside diameter. At the top of column, a double pipe condenser is connected, which is used to condense the vapor leaving the top of column. Part of condensate returns back to the distillation column and the other was drained with a constant reflux ratio Figure (1).

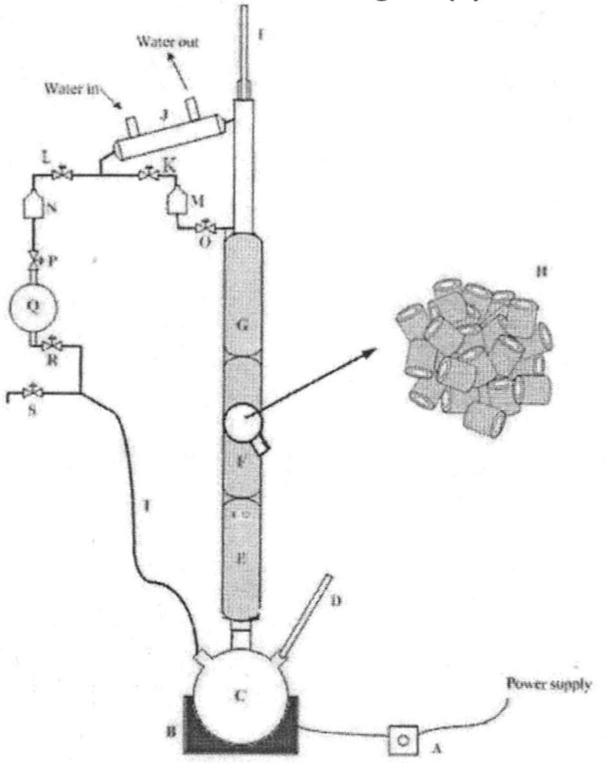

Figure 1 . Schematic Sketch of RD Unit

A: Heat duty controller; **B**: Heating mantle; **C**: Still pot; **D**: Bottom thermometer, **E**: Lower distillation column section; **F**: Intermediate distillation column section **G**: Upper distillation column section; **H**: Rasching ring; **I**: Top thermometer **J**: Condenser; **K**, **L**, **O**, **P**: Reflux valves; **M**: Container; **N**: Container, **Q**: Collecting vessel; **R**: Valve; **S**: Valve; **T**: Recycle line

Measured amounts of reactants were introduced to the still pot together with a desired amount of catalyst, $KHSO_4$, or H_2SO_4 (0.04 kg catalyst/1kg bulk reactant). Cooling water of condenser is started and heat to the still pot is supplied. The reaction takes place in the still between ethanol and tert-butanol then ethyl tert-butyl ether and water produced. Water have the highest boiling point among the still contents, therefore, during the distillation water accumulates in the still and little amount reaches the condenser. Samples were taken from the distillate each hour starting at the second hour because the amount of ETBE produced in the first hour is very small [4, 5, 6 and 23]. Many variables have been studied such as the feed molar ratio (FMR) of 1:1, 2:1, and 4:1, reflux ratio (Rr) of 3, 4, and 5, heat duty of (65, 90, and 146 W), catalyst amount for $KHSO_4$ catalyst, and amount of water in the feed.

Mathematical Model

Consider the batch packed reactive distillation column, the schematic model of jth segment shown in Figures (2).

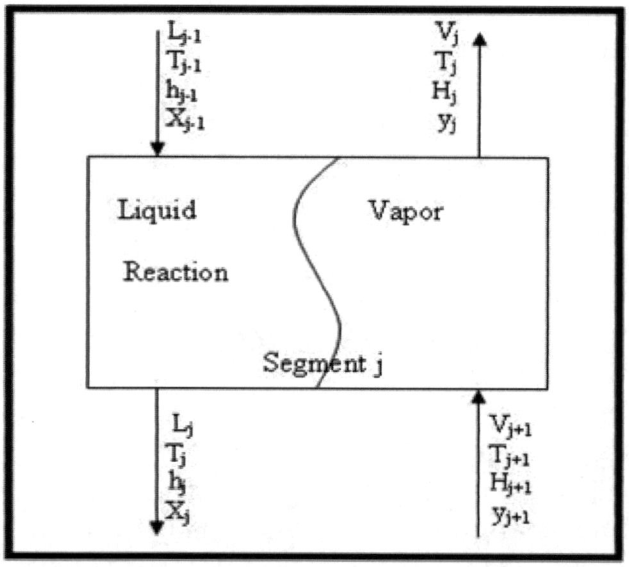

Figure 2. Schematic Diagram of equilibrium Segment

Equations that model the equilibrium segment are shown as MESHR equations. MESHR is an acronym referring to the different types of equations.

M. Total and component material balances.

$$\frac{dM_j}{dt} = V_{j+1} + L_{j-1} - V_j - L_j + R_j \qquad (1)$$

$$\frac{dM_j x_{ij}}{dt} = V_{j+1} y_{i,j+1} + L_{j-1} x_{i,j-1} - V_j y_{i,j} - L_j x_{i,j} + R_{i,j} \quad (2)$$

E. Equilibrium relation

$$y_{i,j} = K_{i,j} x_{i,j} \quad (3)$$

S. Summation equations

$$\sum_{i=1}^{c} x_{i,j} = 1, \sum_{i=1}^{c} y_{i,j} = 1 \quad (4)$$

H. Enthalpy equation

$$\frac{dM_j H_j}{dt} = V_{j+1} H_{j+1} + L_{j-1} h_{j-1} - V_j H_j - L_j h_j + R_j Hr \quad (5)$$

R. Reaction rate equations [23].

$$R = M_{cat} \frac{K_1 (a_A a_B - \frac{a_C a_D}{K_{eq}})}{1 + K_w a_D} \quad (6)$$

$$K_1 = \exp(3.55 - \frac{2286}{T}) \quad (7)$$

$$K_W = \exp(-16.16 - \frac{6636}{T}) \quad (8)$$

$$K_{eq} = \exp(-2101.93 - \frac{38672.73}{T} + 411.41 \log T - 1.592T + 1.345*10^{-3} T^2 - 5.7517*10^{-7} T^3) \quad (9)$$

The activity of i-th component was calculated using the following equation:

$$a_i = x_i \gamma_i \quad (10)$$

- *Enthalpy Calculation*

Enthalpy of component in liquid phase can be estimated from the heat of formation at reference temperature and the sensible heat at a desired temperature [18, 23].

$$h_i = \Delta H_{fi}^{298} + \int_{298}^{T} Cp^L dt \quad (11)$$

Evaluation of integral in Equation (11) requires knowledge of the temperature dependence of heat capacity.

$$Cp_i^L = a + bT + cT^2 + dT^{-2} \quad (12)$$

The total enthalpy of liquid phase given by:

$$h_{i,j} = \sum_{i=1}^{c} x_{i,j} h_i + H_{mix} \quad (13)$$

Where H_{mix} is the heat of mixing, for UNIFAC model heat of mixing is:

$$H_{mix} = RT \sum_{i=1}^{c} (x_i \ln \gamma_i) \quad (14)$$

The enthalpy of component in vapor phase was estimated through the integral of heat capacity in vapor phase at any temperature T with respect to a reference temperature T^o:

$$H_V = \int_{T^\circ}^{T} Cp_V dt \quad (15)$$

Heat capacity of components in the vapor phase is:

$$Cp_i^V = a_V + b_V T + c_V T^2 + d_V T^{-2} \quad (16)$$

Total enthalpy of vapor phase is:

$$H_{V\,i,j} = y_{i,j} H_i \quad (17)$$

Heat of reaction at any temperature can be calculated from heat capacity data if the value for one temperature is known, the tabulation of data can be reduced to the completion of standard heats of formation at a single temperature [21]. The calculation of standard heats of reaction has been given by:

$$Hr^\circ = \sum_{i=1}^{c} v_i H_{fi} \quad (18)$$

Where the sign of stoichiometric ratio v is positive for products and negative for reactants.

$$Hr = Hr^\circ + \int_{T^\circ}^{T} \Delta Cp_i^V dT \quad (19)$$

Enthalpy constants (a_V, b_V, c_V and d_V) for vapor and (*a, b, c* and *d*) for liquid are available in [23].

-*Vapor Pressure Calculation*

The vapor pressure of each component for the present system was calculated using [12, 22].

$$\ln P^{sat} = A1 + \frac{A2}{T + A3} A4 * T + A5 \ln T + A6 * T^{A7} \quad (20)$$

The constants are available in [23].

Where vapor pressure P^{sat} in Pa and T in Kelvin.

-Bubble Point Calculation

Temperatures of stages have been calculated using iterative procedure of bubble point until the summation in Equation (21) equals to one.

$$\sum_{i=1}^{m}(K_{ij}x_{ij}) = 1 \quad (21)$$

Where K is the distribution coefficient and it can be calculated using:

$$K = \gamma_i \frac{P^{sat}}{P} \quad (22)$$

For the present system the activity coefficient γ_i has been calculated using UNIFAC method.

-Holdup

The liquid holdup in a packed column is defined as the volume of liquid held under operating conditions per volume of packed bed. This holdup can be divided into two components, the static and the dynamic (or operating) holdup [19]. The hold up correlations is given by [20,23]:

$$M = \frac{1}{12\varepsilon}(\frac{\mu_L}{\rho_L})^{1/6}(u_L a_P)^{0.5} \quad (23)$$

The holdup in reboiler based on the initial charge to the reboiler (M°) and it is given by [21]:

$$M_{Nt+1} = M^{\circ}_{Nt+1} - \sum_{j=0}^{Nt} M_j - \int_0^t D_t dt \quad (24)$$

Solution of the Equilibrium Model

A computer program to solve the MESHR equations has been developed using MATLAB (R2009b) to determine the composition of components, stages temperatures, condenser and reboiler duties, liquid and vapor flow rates along stages, and reaction rate profile.

The program begins with specify all parameters consist of number of stages, reflux ratio, total pressure, feed compositions, feed flow rate, distillate rate, batch time, step time, and mass of catalyst as well as all physical properties of components. Time, and temperature loops were started, respectively over all stages. The temperature of each stage has been calculated by trail and error until the equilibrium relation applicable.

The new stages temperatures have been used in calculation of reaction rate, enthalpies of vapor, liquid and mixing. Then liquid and vapor flow rates were calculated by total material and energy balances. A tridiagonal matrix was used to find the component compositions by solving the MESHR equations, solving the matrices by eigen value, and normalizing the new compositions for each component. New sets of composition are obtained with the previous procedure for each step time of the batch time. When the compositions at different times are evaluated the program ended and the results plotted.

Results and Discussion

The validity and accuracy of the proposed unsteady state equilibrium model for production of ETBE from ethanol and tert-butanol using MATLAB (R2009b) is checked with Aspen Plus (10.2) program using a rigorous distillation model (RADFRAC) with the same operating conditions in both programs Table(3).

Results of liquid mole fraction in the distillate and bottom with time of the two programs are shown on the same plot for comparison, Figures (3) and (4). ETBE formed is mostly collected in the distillate because it has the lower boiling point 73C^{o} compared with 87, 83, and 100 C^{o} for ethanol, tert-butanol and water, respectively.

The comparison of the two programs shows good agreement in composition profiles with small deviation which is due to the lack of knowledge in the Aspen Plus data of activity coefficients, Antoine parameters, enthalpy of vapor and liquid coefficients, reaction rate kinetics and amount of catalyst used. The error between the two programs which is bounded with $\pm$ 10% lines Figure (5).

Different variables have been studied such as feed molar ratio, reflux ratio, and heat duty and the results were compared with the present experimental work.

-Feed Molar Ratio (FMR)

Experiments 1, 2, and 3 in Table (4) were performed to study the effect of changing FMR on purity of ETBE in the distillate. The observed purity of ETBE in the distillate after 3hrs is 5.619 %mol. Figure (6) shows the effect of changing FMR on the purity of ETBE, as the ratio of ethanol to tert-butanol increases the purity of ETBE decreases this is because the distillate contains the unreacted ethanol which reduces the amount of ETBE in the distillate and the best FMR obtained is 1:1.

- Effect of Changing Reflux Ratio (Rr)

Experiments 1, 2, and 3 in Table (5) were carried out to find the best reflux ratio, for the best value of FMR of 1:1. The purity of ETBE in the distillate when determining the best FMR was low, therefore, the percentage amount of water in the feed is reduced from 95% to 90% and the time increased to 6 hrs. Figures (7) to (9) show the results of liquid composition profile with time in the distillate for experimental and theoretical equilibrium model for reflux ratios of 3, 4 and 5 respectively. (Solid Lines$\equiv$EQ Model, Dashed Lines$\equiv$Experimental).

These Figures show that the purity of ETBE increases with increasing the reflux ratio and the best reflux ratio which gives higher purity of ETBE is 5. This is because when reflux ratio increases the residence time of reactants increased and the concentration of reactants in the distillate reduced therefore, the purity increased.

- Effect of Changing the Heat Duty

All the previous experiments have been carried out under heat duty of 146 W, which is the maximum operating value of the heating mantle. In order to find the effect of changing the heat duty on the purity of ETBE in the distillate; experiments with the best FMR (1:1), best reflux ratio (5), and heat duty of 65 and 90 W were performed. Figures (10) and (11) show the comparison in heat duty between experimental and theoretical model results.

Increasing the heat duty the purity of ETBE increased and the best heat duty is 146 W. When the heat duty increases the vapor rate is increased which transfers more volatile and lower boiling point component from the still to the distillate, therefore, the purity of ETBE which is the lower boiling point component in the system increases.

- Effect of Changing the Amount of Water in Feed

An experiment with 0% water in the feed for catalysts $KHSO_4$ was carried out. Figure (12) shows the comparison between experimental and theoretical model where no water used in the feed for 0.04 kg catalyst/kg reactant.

From the above Figure, the purity of ETBE increases when no water used in feed, and there is amount of tert-butanol (which is the limiting reactant) still unreacted. Higher amount of $KHSO_4$ catalyst are required to increase the purity of ETBE. Therefore, an experiment with higher amount of $KHSO_4$ has been performed.

- Effect of Changing the Catalyst Amount

$KHSO_4$ catalyst has been used with three times its amount in the previous experiments in order to increase the purity of ETBE. Figure (13) shows the liquid mole fraction profiles using feed molar ratio of 1:1, reflux ratio 5, heat duty 146W, and 0.12 kg catalyst/1kg bulk reactant.

From Figure (13) it is clear that the purity of ETBE and conversion of TBA increase with increasing the amount of $KHSO_4$. This is due to more tert-butanol and ethanol reacts to produce ETBE.

- *Effect of Changing the Catalyst Type*

Other catalyst has been chosen such as H_2SO_4 with the same amount (0.04 kg catalyst/1kg bulk reactant), for both 90% and 0% water in the feed, with the best FMR (1:1), best reflux ratio (5), and best heat duty (146W), Figures (14) and (15). The results show when sulfuric acid (H_2SO_4) is used as a catalyst the purity of ETBE is higher than that when potassium hydrogen sulfate is used. For 90% water in the feed and $KHSO_4$ catalyst, the purity of ETBE in the distillate is (0.1487) compared with (0.16615) for sulfuric acid, and for 0% water in the feed the purity is (0.29996) for potassium hydrogen sulfate compared with (0.61317) for sulfuric acid. In the experiments where H_2SO_4 is used the tert-butanol is almost consumed, while when using potassium hydrogen sulfate with the same batch time amount of tert-butanol is still unreacted even it is higher than the amount of ETBE when using 90% water and nearly equal to the amount of ETBE when no water is used for the same amount of catalyst used, this is due to that sulfuric acid is a strong acid causes the tert-butanol and ethanol to reacted faster and therefore, increases the purity of ETBE and conversion of tert-butanol, so for 0% water in the feed the conversion is (85.5775%) when using potassium hydrogen sulfate compared with (94.789%) for sulfuric acid, Figures (16) and (17).

- *Effect of Using Intermediate Reactive Section*

The reaction section has been changed to an intermediate section, where the catalyst $KHSO_4$ is placed in the middle section and the reactants have been introduced to the still, samples were taken every hour for 6h starting from the second hour. The operating conditions of the present experiment are feed molar ratio 1:1, reflux ratio 5, and boiler heat duty 146W, with 160 g amount of catalyst. Figure (18) show that the purity of ETBE in the distillate increased to (0.79407) and the conversion becomes 99.465%, when intermediate reactive section has been used. This is due to the availability of large amount of catalyst.

All above figures show a good agreement between experimental and theoretical results within $\pm 10\%$ errors. Such deviations may be attributed to the little difference in the size and shape of packings and the manual operation of the distillation column when the reflux ratio and heat duty were being set.

Conclusion

As the feed molar ratio of ethanol to tert-butanol increases, the purity of ethyl tert-butyl ether in the distillate decreases.

For the reflux ratio studied in the present work (3, 4, and 5), increasing the reflux ratio cause the purity of ETBE in the distillate to increase.

Changing the heat duty of still affect the composition where the increase of heat duty increases the purity of ETBE at the top.

For the same amount of catalyst used and the same operating conditions (FMR, reflux ratio, and heat duty). When sulfuric acid is used as a catalyst purity of ETBE in the distillate and TBA conversion is higher than that when the potassium hydrogen sulfate is used as a catalyst.

The purity of ETBE can be improved by increasing the amount of catalyst and by using intermediate reactive section, but it is also leads to augments the cost of the process.

References

1. Kenig, E. Y., Gorak, A., Pahalahti,A., Jakobsson, K., ittamaa, J. A, and Sunmacher, K. (2004) "Advanced rate-based simulation tool for reactive distillation", AIChE Journal, Vol. 50, No. 2, 322-342, February.

2. Ozbay, N., and Oktar, N. (2009) "Thermodynamic study of liquid phase synthesis of ethyl tert-butyl ether using tert-butanol and ethanol", J. Chem. Eng. Data, 54, 3208-3214.

3. Al-Arfaj, M. A. and Luyben, W. L., (2002) "Control study of ethyl tert-butyl ether reactive distillation", Ind. Eng. Chem. Res., 41, 3784-3796.

4. Assabumrungrat, S., Wangwattanasate, D., Pavarajarau, V., Praserthdam, P., Arpornwichanop, A., and Goto, S. (2004) "Production of ethyl tert-butyl ether from tert-butanol and ethanol catalyzed by β-Zeolite in reactive distillation", Korean J. Chem. Eng., 21(6), 1139-1146.

5. Umar, M., Saleemi, A. R., and Saha, B. (2009a) "Liquid phase synthesis of ETBE from ethanol and tert-butyl alcohol using macroporous and gelular ion exchange resin catalyst", The Nucleus, 46 (3), 101-107.

6. Umar, M., Saleemi, R., Feroz, N., and Qaiser, S. (2008) "Synthsis of ETBE in packed reactive distillation column using macroporous ion exchange resin catalyst", World Congress on Engineering and Computer Science (WCECS), San Francisco, USA, 22-24 October.

7. Mujtaba I. M., (2004) "Batch distillation design and operation", Series on chemical engineering, Vol. 3.

8. Matouq M., Quitain, A. T., Takahashi, K., and Goto, S. (1996) "Reactive distillation for synthesizing ethyl tert-butyl ether from low grade alcohol catalyzed by potassium hydrogen sulfate", Ind. Eng. Chem. Res., 35,982-984.

9. Sneesby, M. G., Tade, M., Datta, R., and Smith,T. N. (1997b) "ETBE synthesis via reactive distillation 1. Dynamic simulation and control aspects", Ind. Eng. Chem. Res., 36, 1870-1881.

10. Quitain, A. T., H. Itoh, and S. Goto (1999b) "Reactive distillation for synthesizing ethyl tert-butyl ether from bioethanol", Journal of Chemical Engineering of Japan, Vol. 3, No. 3, 280-287.

11. Bisowarno B. H., and Tade, M. O. (2000) "Dynamic simulation of startup in ethyl tert-butyl ether reactive distillation with input multiplicity", Ind. Eng. Chem. Res., 39, 1950-1954.

12. Yang, B., Yang, S., and Wang, H. (2001) "Simulation for the reactive distillation process to synthesize ethyl tert-butyl ether", J. Chem. Eng. of Japan, 34, 1165-1170.

13. Jhon, Y. H., and Lee, T. (2003) "Dynamic simulation for reactive distillation with ETBE synthesis", Separation and Purification Technology, 31, 301-317.

14. Varsili, D., and Dogn, T. (2005) "Simultaneous production of tert-amyl ethyl ether and tert-amyl alcohol from isoamylene-ethanol-water mixtures in a batch reactive distillation column", Ind. Eng. Chem. Res., 44, 5227-5232.

15. Hamza, T. M., and Haweel, C. K. (2005) "Study of performance of batch reactive distillation", M. Sc. Thesis, University of Baghdad, Baghdad, Iraq.

16. Umar, M., Patel, D., and Saha, B. (2009b) "Kinetic studies of liquid phase ethyl tert-butyl ether (ETBE) synthesis using macroporous and gelular ion exchange resin catalyst", Chem. Eng. Sci., 46, 4424-4432.

17. Nakkash, N. B., Al-Habobi, N., and Hamodee, H. M. (2010) "Simulation of batch reactive distillation for methyl acetate production", Spring National Meeting AIChE, T8000. Advances in distillation and absorption, 23-March.

18. Perry, R. H., and Green, D. W. (1997) "Perry's chemical engineer's handbook", McGraw-Hill Companies, Inc.

19. Wagner, I., Stichlmair, J., and Fair, J. R. (1997) "Mass transfer in beds of modern, high efficiency random packings", Ind. Eng. Chem. Res., 36, 227-237.

20. Kister, H. Z. (1992)"Distillation design", McGraw-Hill, Inc., New York .

21. Seader, J.D., and Henley, E. J. (1998) "Separation process principles", John Wiley and Sons, Inc., New York.

22. Walas, S. M. (1985) "Phase equilibria in chemical engineering", Butterworth Publications, London.
23. Al-Janabi, N. K. (2010) "Simulation and Experimental Investigation to produce Ethyl Tert-Butyl Ether using Reactive Distillation", M.Sc. Thesis, Al-Nahrain University, Baghdad, Iraq.

Notations

Symbols	Notations
A1,A2,A3,A4 ,A5	Antoine Equation Constants
a_i	Activity
A, b, c, d	Constants of Heat Capacity of Liquid Phase
a_v, b_v, c_v, d_v	Constants of Heat Capacity of Vapor Phase
Cp	Specific Heat
E	Energy
H_v	Enthalpy of Vapor Phase
H	Enthalpy of Liquid Phase
H_{mix}	Heat of Mixture
Hr°	Reference Heat of Reaction
ΔH_{vb}	Heat of Vaporization at Normal Boiling Point
Hr	Heat of Reaction
h^V	Heat Transfer Coefficient in Vapor Phase
h^L	Heat Transfer Coefficient in Liquid Phase
K	Equilibrium Constant
L	Liquid Flow Rate
M_{cat}	Mass of Catalyst
M	Liquid Hold-Up
N	Total Number of Component
P	Pressure
P^{sat}	Saturated Pressure
R_i	Reaction Rate
T^o	Reference Temperature
V	Vapor Flow Rate
X	Liquid Mole Fraction
Y	Vapor Mole Fraction

Greek Letters

γ	Activity Coefficient

Subscripts

i, j	Component indices
L	Liquid
V	Vapor
N_t	Total Number of Stages
N_{t+1}	Reboiler

Superscripts

Sat	Saturated
L	Liquid
V	Vapor

Abbreviations

RD	Reactive Distillation
ETBE	Ethyl Tert-Butyl Ether
MTBE	Methyl Tert-Butyl Ether
FMR	Feed Molar Ratio
IB	Isobutene
ETOH	Ethanol
TBA	Tert-Butanol
FVR	Feed Volume Ratio
Rr	Reflux Ratio

Table 3. Operating Conditions for Comparison of Proposed EQ Program and Aspen Plus [4].

Pressure (Pa)	101325
Feed molar ratio (TBA:ETOH:H_2O)	1:1:38
Composition [% mol]	
ETOH	2.5
TBA	2.5
H_2O	95
Feed flow rate (mol/s)	$2.7*10^{-3}$
Reflux ratio	1.5
Catalyst weight per stage (kg)	0.04
Total stages	16
Rectification stages	5
Reaction stages	6
Stripping stages	5
Feed stage	12
Heat duty (W)	26.3

Table 4. Studied Operating Conditions for Changing FMR

No.	Reflux ratio	FMR	Heat duty	Initial water content	Catalyst used
1	Total	1:1	146 W	95%	$KHSO_4$
2	Total	2:1	146 W	95%	$KHSO_4$
3	Total	4:1	146 W	95%	$KHSO_4$

Table 5. Studied Operating Conditions for the Changing the Reflux Ratio

No	Reflux ratio	FMR	Heat duty	Initial water content	Catalyst used
1	5	1:1	146 W	90%	$KHSO_4$
2	4	1:1	146 W	90%	$KHSO_4$
3	3	1:1	146 W	90%	$KHSO_4$

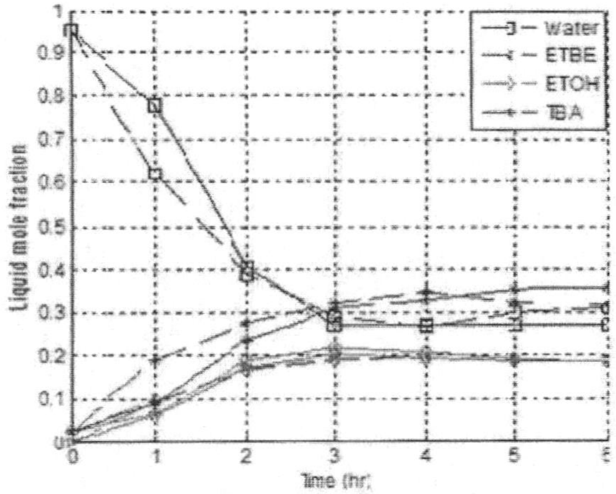

Figure 3. Comparison of the Proposed Program and Aspen Plus, Distillate, Solid lines: ASPEN PLUS, Dashed Lines: Proposed Program

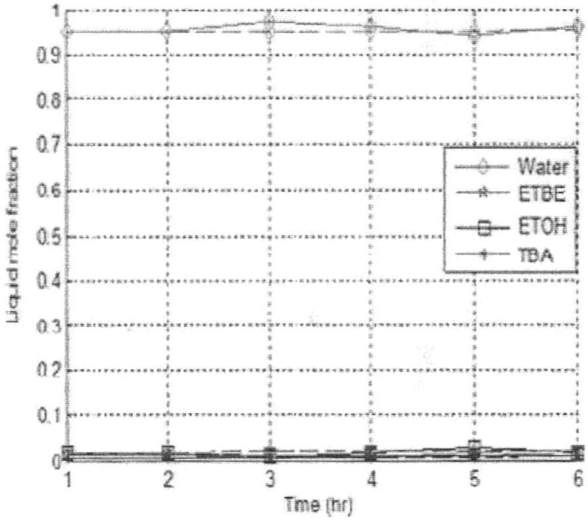

Figure 4. Comparison of the Proposed Program and Aspen Plus, Bottom, Solid lines: ASPEN PLUS, Dashed Lines: Proposed Program

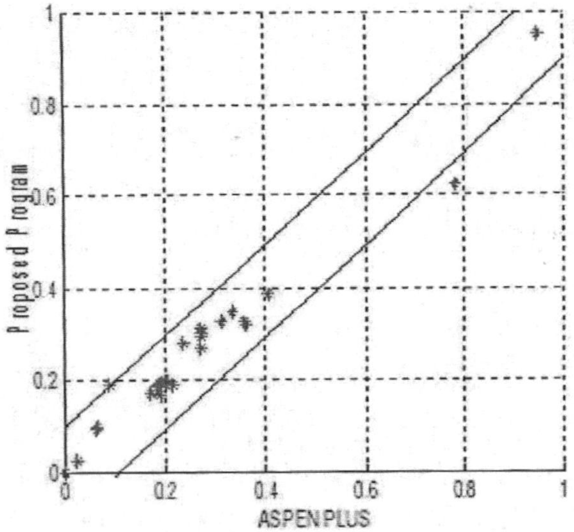

Figure 5. Error between the Two Programs in the Distillate

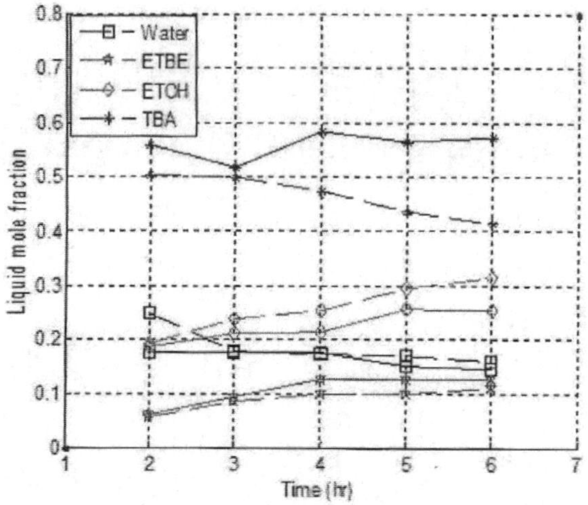

Figure 7. Experimental and EQ Model Composition profile in the Distillate for Rr=3, Heat Duty=146W

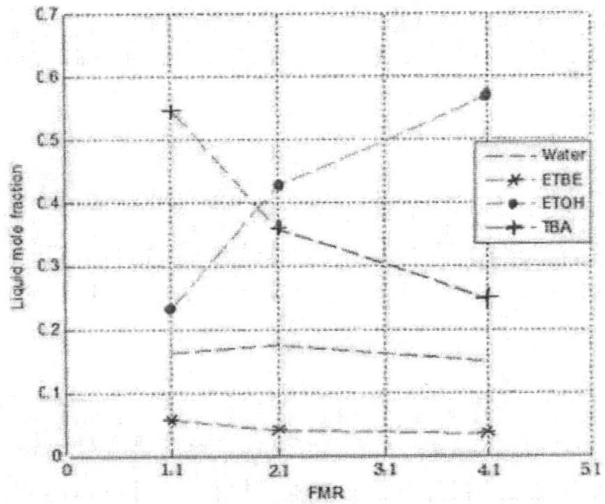

Figure 6. Liquid Mole Fractions with Different FMR

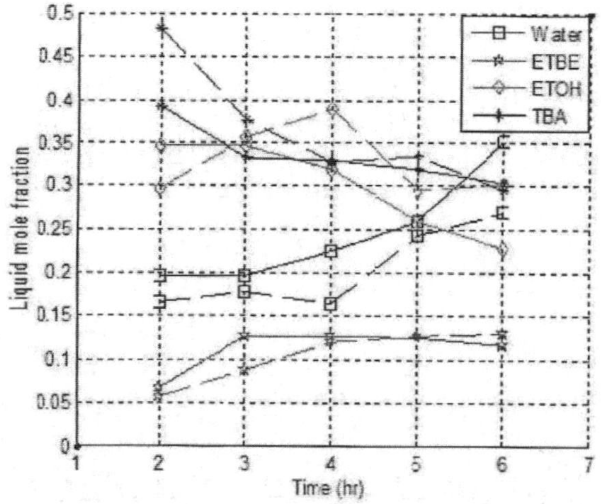

Figure 8. Solid Lines≡EQ, Dashed Lines≡Experimental, Rr=4, Heat Duty=146W

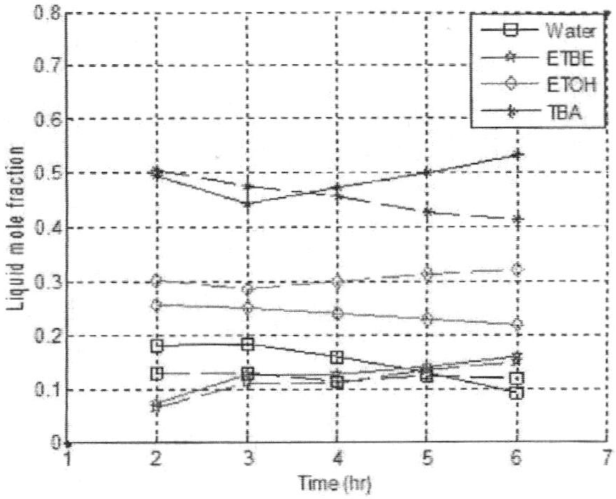

Figure 9. Solid Lines ≡ EQ, Dashed Lines ≡ Experimental, Rr=5, Heat Duty=146W

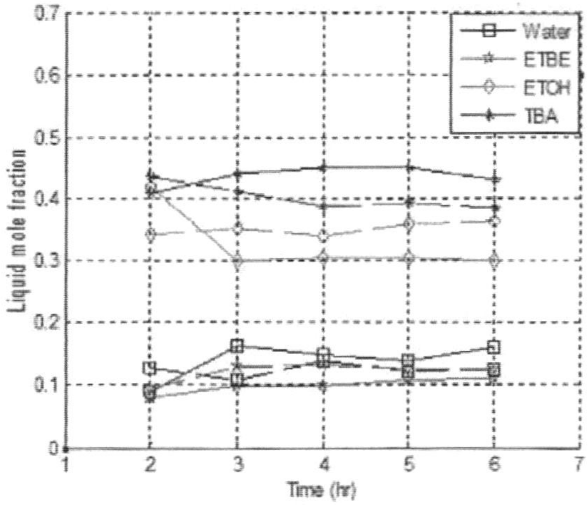

Figure 10. Solid Lines ≡ EQ, Dashed Lines ≡ Experimental, Heat Duty=65W

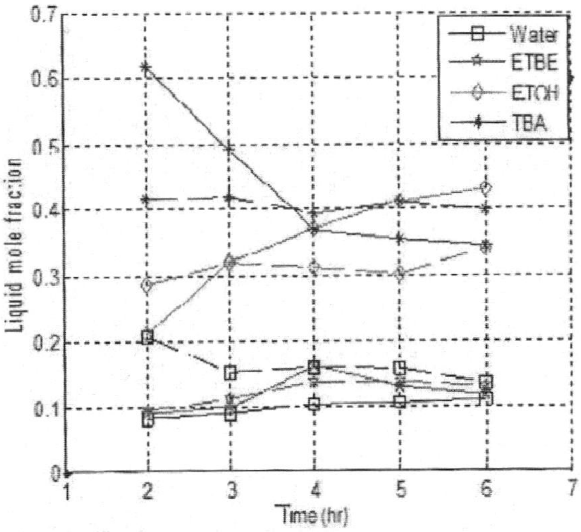

Figure 11. Solid Lines ≡ EQ, Dashed Lines ≡ Experimental, Heat Duty=90W

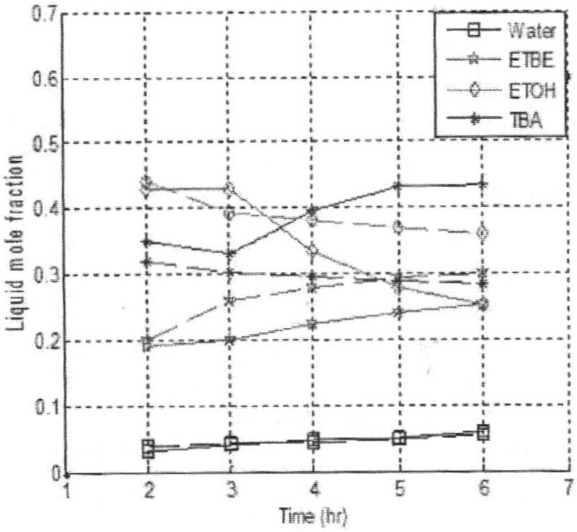

Figure 12. Solid Lines ≡ EQ, Dashed Lines ≡ Experimental, 0%Water in Feed

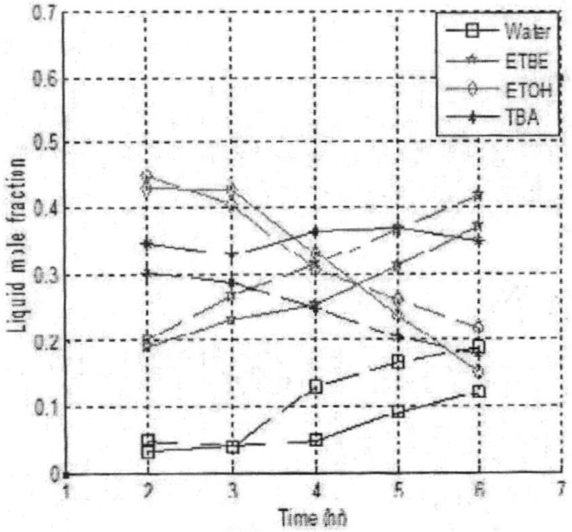

Figure 13. Solid Lines ≡ EQ, Dashed Lines ≡ Experimental, 0%Water in Feed with Three Times Catalyst $KHSO_4$

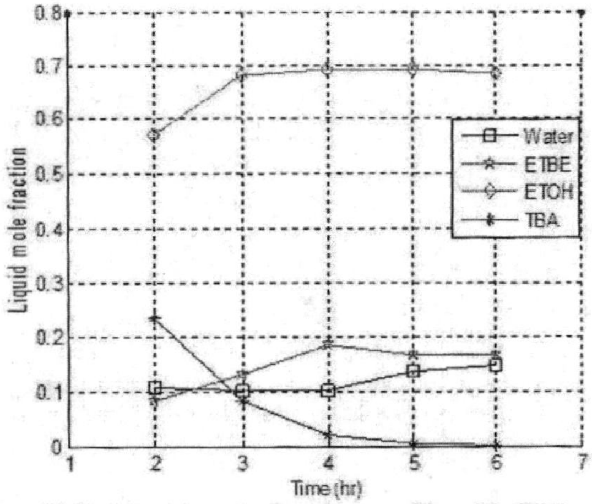

Figure 14. Liquid mole fraction profile with 90% water and catalyst H_2SO_4

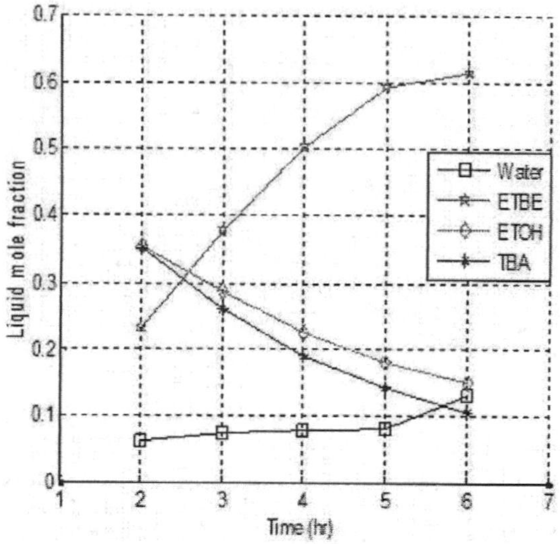

Figure 15. Liquid mole fraction profile with 0% water and catalyst H_2SO_4

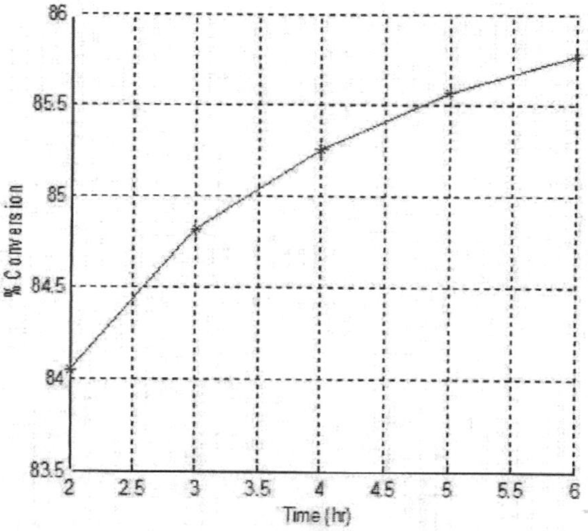

Figure 16. Conversion profile for 0% water and catalyst $KHSO_4$

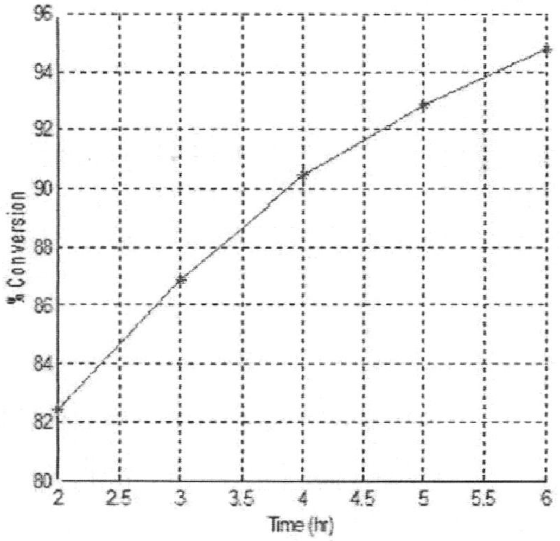

Figure 17. Conversion profile for 0% water and catalyst H_2SO_4

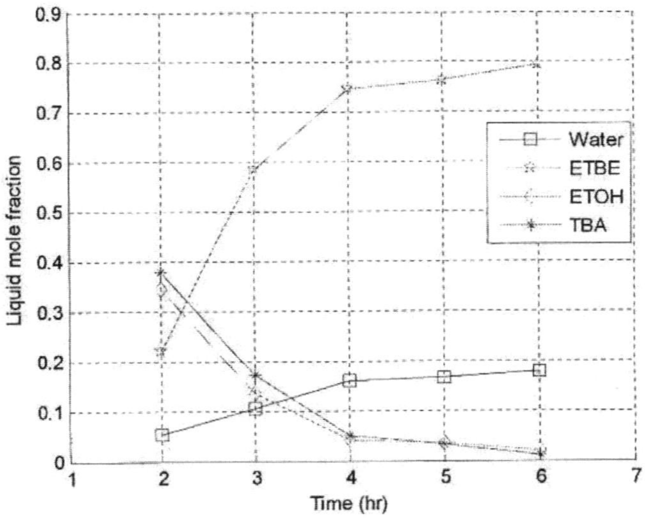

Figure 18: Liquid mole fraction profile using intermediate reactive section

SIMULATION OF CONTINOUS REACTIVE DISTILLATION OF ETBE PRODUCTION FROM ISOBULENE

Nada Bahjat Nakkash, Nadeen Khalid Al-Janabi
Al-Nahrain University, Baghdad – Iraq

Abstract

The present work concerned with simulating a continuous multicomponent reactive distillation (RD) using equilibrium model for the production of ethyl tert-butyl ether from isobutene and ethanol. A simulation model was performed for RD column with ETBE synthesis using MATLAB (R2009b) to solve MESHR equations.

The developed simulation model was checked and compared with a previous work and RADFRAC model of ASPEN PLUS 10.2, the results gave a good agreement and with $\pm 10\%$ error.

Split feed condition was considered and the influence of various parameters on the performance and behavior of RD for ETBE production, such as reflux ratios in range of 2 to 8, mass of catalyst of 60 to 190 g, number of reaction stages of 3 and 4 has been studied.

The best design conditions for operating RD distillation column to obtain high purity of ETBE 89.89% with 40.7% selectivity and 99.82% conversion was at reflux ratio 5, number of stages 10, and mass of catalyst 120g per stage using single feed. When split feed is considered with ETBE fed at stage 2 and ETOH, IB and n-B at stage 6 and when ETBE and ETOH fed at stage 2, IB and n-B at stage 6 the best design conditions was at reflux ratio 5, number of stages 10, mass of catalyst 90g per stage and three reactive stages.

Keywords: *Continuous Reactive Distillation, Simulation, ETBE Production*

Introduction

Reactive distillation involves combined simultaneous chemical reaction and distillation. The chemical reaction usually takes place in the liquid phase or at the surface of a solid catalyst in contact with the liquid phase [1]. The use of reactive distillation has grown because it results in less expensive and more efficient processes for some chemical systems. There is increasing interest in the use of ethyl tert-butyl ether (ETBE) for gasoline blending as a replacement for methyl tert-butyl ether (MTBE) because of the latter's environmental problems [2].

The basic reaction combines ethanol and isobutene to form ETBE.

$$iC_4^{=} + Ethanol \Leftrightarrow ETBE$$

The isobutene feed stream contains both isobutene and normal butene, and the latter does not participate in the reaction. The chemically inert normal butene is much lighter than ETBE, so the distillate is mostly n-butene and the bottom is mostly ETBE. Any unreacted isobutene leaves mostly in the distillate. If an excess of ethanol is fed to the column, this ethanol will typically leave in the bottoms and be blended with ETBE in gasoline pool [2, 3].

Sneesby et al. [3] presented continuous mathematical models for reactive distillation of ETBE synthesis from isobutene and ethanol using both Pro II and Speed Up simulation software. Comparison between the two programs shows excellent agreement with each other.

Quitain et al. [4] proposed an industrial scale process for synthesis of ETBE from tert-butanol and simulated using ASPEN PLUS. The overall conversion of TBA and the overall ETBE selectivity are 98.9% and 99.9% respectively.

Al-Arfaj and Luyben[5] explored the effect of number of trays in the rectifying and stripping sections of reactive distillation columns for ETBE and a hypothetical system. This work demonstrates that adding additional trays in a reactive distillation column does not degrade performance, provided the specified degrees of freedom are appropriately chosen.

Tade and Tian [6] showed that the conversion could be inferred from multiple process temperatures. For 10-stages ETBE reactive distillation pilot plant process with six temperature measurements, a third-order nonlinear inferential model was developed using MATLAB software to infer the reactant conversion with sufficiently high accuracy. The two temperatures employed in the model were the bottom reactive section temperature and the reboiler temperature, respectively.

Al-Arfaj and Luyben [2] explored two process configurations for ETBE synthesis: a design with two fresh reactant feed streams and a design with a single mixed reactant feed. Several control structures are investigated. Their results showed that the double feed system requires internal composition control to balance the stoichiometry along with temperature control to maintain product purity, while, the single feed case is effectively controlled with only a temperature controller.

Jhon et al., [7] presented a dynamic simulation model using MATLAB software for reactive distillation column of 20 stages with ETBE synthesis. A structured and simple iterative algorithm was devised to estimate the vapor flows with the fast convergence under the rigorous energy balance. Both total and partial condensers are considered, and the reactive distillation performance for these condensers was observed. By comparison of their influence on purity of ETBE, total condenser is the better one.

Yanghong et al. [8] studied the variation of reflux ratio, reboiler rate, and feed molar ratio of ETOH/IB and their influence on the conversion of IB on ETBE purity. Experimental results showed that higher conversion of IB was obtained with feed molar ratio of 1 to 1.1, reflux ratio 6, and 0.115 m^3/hr reboiler rates. For these conditions conversion was exceeded 90%.

Bisowarno et al. [9] presented a mathematical design model of single and double feed reactive distillation column for ETBE production and investigated the effects of separation and reaction stages on the overall performance. Their results showed that for the single feed column, longer column requires both reflux ratio and reboiler duty to be adjusted to optimize the ETBE purity, while the increasing of reflux ratio does not reduce the ETBE purity for the shorter column. For the double feed column, increasing reflux ratio does not degrade the column performance.

Bolun et al. [10] modeled a process for synthesis of ETBE with ASPEN PLUS simulation package. They were observed the input multiplicity for a range of reboiler duty of several values of reflux ratio. These results can be used to avoid the excessive energy consumption and achieve optimum design of reactive distillation column. They found that the best reflux ratio is 5 over a narrow range of reboiler duty (78-83 kW).

Gomez et al. [11] proposed a mixed integer nonlinear programming (MINLP) formulation for optimal design of a reactive distillation column based on a generic nonequilibrium model. Catalytic distillation of ethyl tert-butyl ether production from isobutene and ethanol is considered as an illustrative example.

Mohammed et al. [12] developed a steady state model for ETBE and MTBE synthesis using MATLAB program. Liquid composition, vapor composition, and temperature profiles were determined. Also, the residue curve maps were plotted for reacted system and system without reaction.

Present Work

The present work emphasized to construct a simulation model for continuous equilibrium reactive distillation of ethyl tert-butyl ether (ETBE) production from isobutene and ethanol over an acidic ion exchange resin, Ambelyst 15.

The profiles of stage temperature, vapor and liquid flow rates, rate of reaction, composition of vapor and liquid were computed using MESHR equations.

The percentage conversion and selectivity are calculated for both single feed and split feed conditions using different amounts of catalyst of 60g to 190g, reflux ratios of 2 to 8 and number of reaction stages 3 and 4.

The initial input conditions and data for the developed simulation model of RD are given in Table (1) for single intermediate feed and the simulation results accuracy were checked with ASPEN PLUS 10.2 software[13] and with previous work [3].

The physical and chemical properties are taken from literature [14, 15] and due to the highly nonideal nature of ether-alcohol-alkenes system. UNIFAC model has been used to predict liquid phase activity coefficient.

In the present work the vapor phase is assumed to be nearly ideal, liquid feed is used, total condenser to give a maximum heat removed at the top of the column, constant molar liquid holdup, neglect the vapor holdup, perfect mixing at each stage.

Mathematical Model

The reactive distillation column was modeled using N-equilibrium stages including total condenser and partial reboiler. Schematic of RD column is shown in Figure (1), and schematic of single stage is in Figure (2).

The following assumptions were made when formulating the reactive distillation process:

1. The vapor and liquid are in equilibrium at each stage.
2. The reaction occurs in the liquid phase.
3. The heat loss and mass transfer resistance can be neglected.

MESH equations and reaction rate were used as the basic equations for the steady state reactive distillation model.

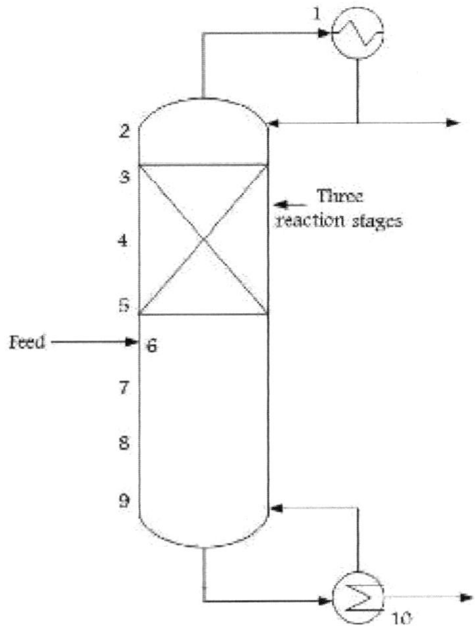

Figure 1. Schematic of RD column

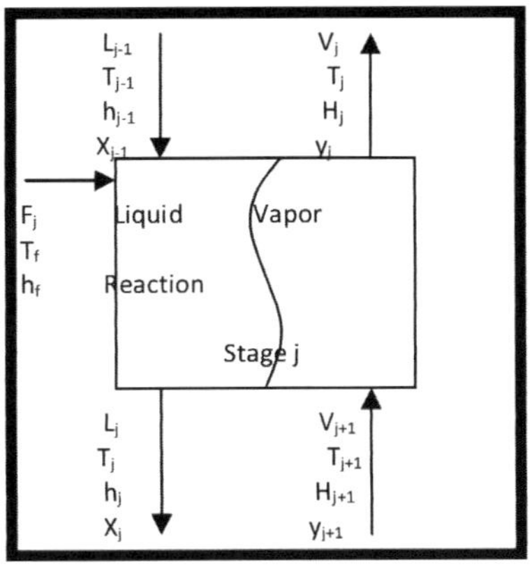

Figure 2. Schematic of a single stage

The MESHR equations are:

-Separation Stage

Total material balance:

$$L_{j-1}+V_{j+1}-V_j-L_j=0 \tag{1}$$

Component balance:

$$V_{j+1}y_{i,j+1}+L_{j-1}X_{i,j-1}-V_jy_{i,j}-L_jX_{i,j}=0 \tag{2}$$

Energy balance:

$$V_{j+1}H_{j+1}+L_{j-1}h_{j-1}-V_jH_j-L_jh_j=0 \tag{3}$$

Phase equilibrium relation:

$$y_{i,j}=K_{i,j}x_{i,j} \tag{4}$$

Summation equations:

$$\sum_{i=1}^{c}X_{i,j}=1,\sum_{i=1}^{c}y_{i,j}=1 \tag{5}$$

-Reactive Stage

Total material balance:

$$L_{j-1}+V_{j+1}-V_j-L_j+R_j=0 \tag{6}$$

Component balance:

$$V_{j+1}y_{i,j+1}+L_{j-1}X_{i,j-1}-V_jy_{i,j}-L_jX_{i,j}+r_{ij}=0 \tag{7}$$

Energy balance:

$$V_{j+1}H_{j+1}+L_{j-1}h_{j-1}-V_jH_j-L_jh_j-r_j\Delta Hr=0 \tag{8}$$

-Total Condenser

Total material balance:

$$V_2-L_1=0 \tag{9}$$

Component balance:

$$V_2y_{i,2}-L_1X_{i,1}=0 \tag{10}$$

Energy balance:

$$V_2H_2-L_1h_1=0 \tag{11}$$

-Partial Reboiler

Total material balance:

$$L_{Nt-1}-V_{Nt}-L_{Nt}=0 \tag{12}$$

Component balance:

$$L_{Nt-1}X_{i,Nt-1}-V_{Nt}y_{i,Nt}-L_{Nt}X_{i,Nt}=0 \tag{13}$$

Energy balance:

$$L_{Nt-1}h_{Nt-1}-V_{Nt}H_{Nt}-L_{Nt}h_{Nt}=0 \tag{14}$$

-The Reaction Kinetics

ETBE is produced by the reversible reaction of isobutene and ethanol over an acid catalyst, such as acidic ion-exchange resin, Amberlyst 15.

$$(CH_3)_2C=CH_2+C_2H_5OH\Leftrightarrow(CH_3)_2COC_2H_5$$

The reaction rate expression [2, 3, 5] is as follows:

$$k_{ETBE} = 10.387 + \frac{4060.59}{T} - 2.89055 \ln T$$

$$- 0.0191544\, T + 5.28586 * 10^{-5} T^2 - 5.32977 * 10^{-8} T^3 \quad (15)$$

Generalized rate equation

$$r = \frac{m_{cat} K_{rate} a_{ETOH}^2 (a_{IB} - \frac{a_{ETBE}}{K_{ETBE} a_{ETOH}})}{(1 + K_A a_{ETOH})^3} \quad (16)$$

$$\ln K_A = -1.0707 + \frac{1323.1}{T} \quad (17)$$

$$K_{rate} = 7.418 * 10^{12} \exp(\frac{-60.4}{RT}) \quad (18)$$

ETBE reaction system can also include a side reaction; the dimerization of isobutene to produce di-isobutene:

$$(CH_3)_2C{=}CH_2 + (CH_3)_2C{=}CH_2 \Leftrightarrow [(CH_3)_2C{=}CH_2]_2$$

However, this side reaction is eliminated when ethanol is in excess over the isobutene [2].

If any water is present in the reaction environment, one undesirable reaction (dehydration of isobutene to form isobutyl alcohol) is possible.

$$(CH_3)_2C = CH_2 + H_2O \Leftrightarrow (CH_3)_3COH$$

A computer program to solve the MESHR equations has been developed using MATLAB (R2009b) to determine the composition of vapor and liquid, stages temperatures, liquid and vapor flow rates through the tower, reaction rate at each stage, and condenser and reboiler duties.

The program begins with specify all specifications consist of number of stages, reflux ratio, total pressure, feed compositions, feed flow rate, distillate rate, and mass of catalyst as well as all physical properties of components. Iteration of stage temperature and composition loops was started respectively on each stage. The temperature of each stage has been calculated by trial and error until the equilibrium relation applicable, where the distribution coefficient computed from $K_{i,j} = \gamma_i \frac{P^S}{P}$ for nonideal system. Antoine equation has been used to predict the vapor pressures P^S and the coefficients are taken from literature [14].

$$\ln \frac{P^S}{P_C} = (1-w)^{-1}(Aw + Bw^{1.5} + Cw^3 + Dw^6) \quad (19)$$

Where $$w = 1 - \frac{T}{T_C} \quad (20)$$

The liquid activity coefficient γ_i for the reaction mixture of ETBE synthesis can be computed using UNIFAC method [15]. The equations and parameters to evaluate γ_i for ETBE synthesis are taken from literature [2, 15]. The new stages temperatures have been used in calculation of reaction rate and liquid and vapor flow rates by material and energy balances. For energy balance, total liquid enthalpy H_{Lj} and vapor enthalpy H_{Vj} are calculated and their coefficients were obtained from literature [14].

The vapor enthalpy

$$H_i(j) = h_{fi} + C_{1iv}(T_i(j) - T_o) + C_{2iv}(T_i(j)^2 - T_o^2)/2 + C_{3iv}(T_i(j)^3 - T_o^3)/3 + C_{4iv}(T_i(j)^4 - T_o^4)/4 \quad (21)$$

$$H(j) = \Sigma H_{Vi}(j) * y_{i,j} \quad (22)$$

The liquid enthalpy

$$h_i(j) = c_{1il}(T_i(j) - T_o) + c_{2il}(T_i(j)^2 - T_o^2)/2 + c_{3il}(T_i(j)^3 - T_o^3)/3 + c_{4il}(T_i(j)^4 - T_o^4)/4 \quad (23)$$

$$H_{mix} = RT\, \Sigma\, x_i \ln \gamma_i \quad (24)$$

$$h(j) = \Sigma\, h_i(j) * x_{i,j} + H_{mix} \quad (25)$$

A tridiagonal matrix was used to find the compositions of each component by solving the MESHR equations, solving the matrices by Eigen value, and normalizing the compositions, new values of temperatures, reaction rates, vapor and liquid flow rates on each tray have been calculated. Then the condenser and reboiler heat duties were computed.

Results and Discussions

The proposed model is used to calculate the temperature, liquid and vapor flow rates, reaction rate and composition profiles and the heat input to the still and the heat removed from the condenser for ETBE production from Isobutene and Ethanol. The results of the developed model using initial conditions given in Table (1) are shown in Figure (3). Figure (3) shows a good agreement in composition profile of the developed program and Aspen Plus results with about $\pm$10%error, as in Figure (4).

Deviation of the proposed program and Aspen Plus is expected because Aspen Plus uses built in data such as the data of UNIFAC method, enthalpy, reaction rate equation including the mass of catalyst, and Antoine equation.

The purity of ETBE at the bottom of tower is 89.89% with40.7% selectivity and 99.82% conversion when single feed is used at stage 6, reflux ratio 5, amount of catalyst 120g per stage and three reaction stages, Figure(3). As the amount of catalyst decreased to 110g per stage the purity decreases to 86.1% with 99.75 % conversion and 38.97 % selectivity.

When split feed is considered for the same operating conditions, ETBE fed to stage 2 and IB, ETOH and n-B fed to stage 6 using 100g per stage catalyst. ETBE purity increased to 99.69%, Figures (5). When ETBE and ETOH fed to stage 2 and IB and n-B fed to stage 6 the purity of ETBE, conversion and selectivity stays approximately the same. Thus it is obvious that using split feed the purity of ETBE, conversion and selectivity increases with decreasing the amount of catalyst used.

When the number of stages increased to 4 and the mass of catalyst increased to 190g per stage, the purity of ETBE decreased to 94.85% when ETBE fed to stage 2 and ETOH, IB and n-B fed to stage 6 Figure(6). While ETBE purity becomes 96.33% when ETBE and ETOH fed to stage 2 and IB, and n-B fed to stage 6 Figure(7). Thus, when the reactive stages increased from 3 to 4 the purity of ETBE decreases, this is because increasing the reactive stages increasing ETBE in the lower stages and shifts the chemical equilibrium back to the reactants will decrease the purity of ETBE.

Figure (8, 9, 10, and 11) show the effect of change reflux ratio on conversion and Figures (12, 13, 14 and, 15) show the effect of changing the reflux ratio on selectivity for split feed condition using three and four reactive stages. In all cases increases the reflux ratio caused the conversion to increase where the best reflux ratio is 5, increasing the reflux above the best the conversion remains nearly constant. Also increasing the reflux ratio will increase the selectivity of ETBE, for all cases the best reflux ratio with respect to selectivity is 7. Although increasing the reflux ratio over 5 (which is the best value for conversion) increases the selectivity by nearly 10%, in practical applications operating at higher reflux ratio is not economic due to the requirement of higher heat duty. When the reflux ratio increases the residence time of reactants increased and the concentration of reactants in the distillate decreased and therefore, the conversion and selectivity increased.

For reflux ratio 5 and split feed condition as the mass of catalyst increases there is a slight increase in purity of ETBE, Tables (2, 3, 4, and 5).

Conclusion

1. The simulation model developed gives a good agreement with ASPEN PLUS 10.2 program with about $\pm 10\%$ errors for the same operating conditions
2. Using single feed the purity of ETBE in the bottom is 89.89% for 120g/stage of catalyst, when split feed is used the purity of ETBE increased to 99.69% for 100g/stage catalyst.
3. Increasing the number of reactive stages above three increases ETBE in the lower stages and shifts the chemical equilibrium back to the reactants and therefore decreases the purity of ETBE.
4. Increasing the reflux ratio increase the residence time of reactants and decrease the concentration of reactants in the distillate. Thus, the purity of ETBE increases and the best reflux ratio is 5.
6. Increase the mass of catalyst for reflux ratio 5 and split feed condition cause slight increase in the purity of ETBE.

References

1. Seader J. D. and Henley J. R. (1998), "Separation process principles"; John Wiley and Sons.
2. Al-Arfaj M. A. and Luyben W. L. (2002), "Control study of ethyl tert-butyl ether reactive distillation"; Ind. Eng. Chem. Res.; 41, 3784-3796.
3. Sneesby M. G., Tade M. O., Datta R., and Smith T. N. (1997); "ETBE synthesis via reactive distillation. 1. Steady state simulation and design aspects"; Ind. Eng. Chem. Res.; 36, 1855-1869.
4. Quitain A., Itoh H., and Goto S.(1999); "Industrial scale simulation of proposed process for synthesizing ethyl tert-butyl ether from bioethanol" Journal of chemical engineering of Japan, vol. 32, No. 4, pp. 539-543.
5. Al-Arfaj M. A. and Luyben W. L.(2000); "Effect of number of fractionating trays on reactive distillation performance", AIChE Journal, December, vol. 46, No. 12, 2417-2425,.
6. Tade M. O., and Tian Y.(2000); "Conversion influence for ETBE reactive distillation", Separation and purification technology, 19, 85-91.
7. Jhon Y. H., and Lee T.(2003); "Dynamic simulation for reactive distillation with ETBE synthesis", Separation and purification technology, 31, 301-317.
8. Yanghong Li, Xigang Yuan, Shaomin Wa (2003); "Zeolite film packing application to synthesis of ETBE in a catalytic distillation column, Fuel Chemistry Division Preprints,48,(1),439.
9. Bisowarno B. H., Tade M. O., and Tian Y. (2004); "Interaction of separation and reactive stages on ETBE reactive distillation", AIChE Journal, March, vol. 50, No. 3, 646-653.
10. Bolun Y., Jiang W., Guosheng Z., Huajum W., and Shiqing L. (2006) "Multiplicity analysis in reactive distillation column using ASPEN PLUS", Chinese J. Chem. Eng., 14, 3, 301-308.
11. Gomez J. M., Reneaume J., Roque M., Meyer M., and Meyer X. (2006) " A mixed integer non linear programming formulation for optimal design of a catalytic distillation column based on a generic non equilibrium model", Ind. Eng. Chem. Res.; 45, 1373-1388.
12. Mohammed M. Z., and Shakoor Z. M. (2009)" modeling and simulation of

reactive distillation with MTBE synthesis" M. Sc. Thesis, University of Technology, Baghdad-Iraq.
13. ASPEN Technology Inc. (1993) "The Speed UP User's Manual"; Aspen Technology, Inc: Cambridge, MA.
14. Reid R.C.; Prausnitz,J.M.; Poling B.E. (1987); "The properties of Gases and Liquids", 4th Edt., Mc-Graw Hill; New York.
15. Smith J.M., Vanness, H.C., Abbot M. M. (2001) "Introduction to Chemical Engineering Thermodynamics" sixth edition. Mc-Graw Hill, New York.

Nomenclature

Aw, Bw, Cw, Dw	Antoine coefficients	
A	Activity of species	
c1, c2, c3, c4	Coefficients of liquid enthalpy	
C1, C2, C3, C4	Coefficients of vapor enthalpy	
H	Enthalpy of vapor	kJ/kmol
H_{mix}	Enthalpy of mixing	kJ/kmol
Hr	Heat of reaction	kJ/kmol
H	Enthalpy of liquid	kJ/kmol
K	Distribution coefficient	
K_A	Absorption equilibrium constant	
K_{rate}	Reaction rate constant	mol/hr. g of catalyst
K_{ETBE}	ETBE reaction equilibrium constant	
L	Liquid flow rate	kmol/hr
m_{cat}	Mass of catalyst	G
P	Total pressure	Atm
P^S	Vapor pressure	Bar
P_C	Critical pressure	Bar
R	Gas constant=8.314	J/mol. K
R	Reaction rate	mol/hr
T	Temperature	K
T_o	Reference temperature	K
T_C	Critical temperature	K
V	Vapor flow rate	kmol/hr
X	Liquid mole fraction	
Y	Vapor mole fraction	

Greek letters

Γ	Activity coefficient
Δ	Change of difference

Subscripts

I	Referring to species
J	Referring to stage
1	Condenser,stage1
2	Stage 2
Nt	Reboiler

Table 1. Initial input simulation data [3]

Feed composition	ethanol=0.091; isobutene=0.073; ETBE=0.291; n-butene=0.545;
Total number of stages	10
Reflux ratio	5
Total pressure	9.5 atm
Condenser	Total
Reboiler	Partial
Number of Reaction stages	3
Feed stage location	6
Feed rate	0.407 kmole / hr
Distillate rate	0.217 kmole / hr
Feed temperature	30°C

Table 2. Effect of changing the mass of catalyst on the bottom composition, three reactive stages, ETBE fed in stage 2

Component	Mcat (g per stage)				
	60	70	80	90	100
ETOH	0.0019	0.0019	0.0019	0.0019	0.0001
IB	0.0001	0.0001	0.0001	0.0001	0.001
ETBE	0.9969	0.9969	0.9968	0.997	0.9969
n-B	0.0011	0.0011	0.0012	0.001	0.002

Table 3. Effect of changing the mass of catalyst on the bottom composition, three reactive stages, ETBE and ETOH fed in stage 2

Component	Mcat (g per stage)				
	60	70	80	90	100
ETOH	0.0019	0.0019	0.0019	0.0019	0.002
IB	0.0001	0.0001	0.0001	0.0002	0.0001
ETBE	0.9968	0.9968	0.9967	0.9969	0.9968
n-B	0.0011	0.0012	0.0012	0.001	0.0011

Table 4. Effect of changing the mass of catalyst on the bottom composition, four reactive stages, ETBE fed in stage2

Components	Mcat (g per stage)		
	180	185	190
Ethanol	0.018	0.017	0.0159
IB	0.0015	0.0014	0.0013
ETBE	0.968	0.9699	0.9719
n-B	0.0126	0.0118	0.011

Table 5. Effect of changing the mass of catalyst on the bottom composition, four reactive stages, ETBE and ETOH fed in stage 2

Components	Mcat (g per stage)		
	180	185	190
ETOH	0.04937	0.04872	0.0483
IB	0.002325	0.00222	0.002157
ETBE	0.9455	0.9463	0.9467
n-B	0.0081	0.00774	0.00749

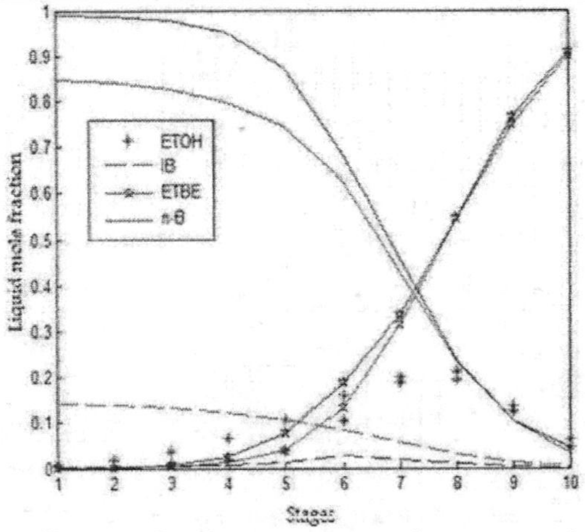

Figure 3. Comparison of proposed program and Aspen Plus

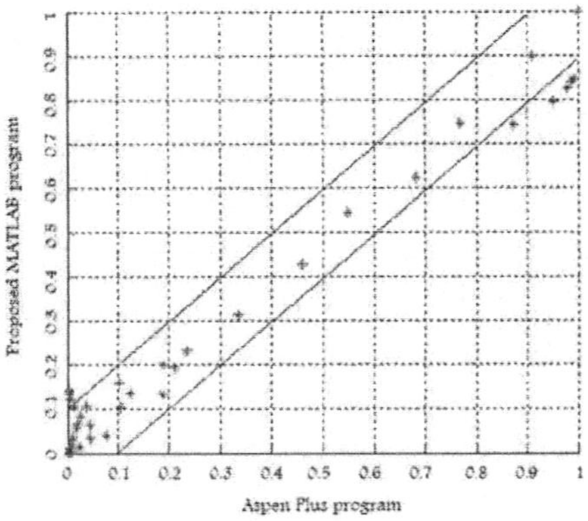

Figure 4. Error of proposed program and Aspen Plus

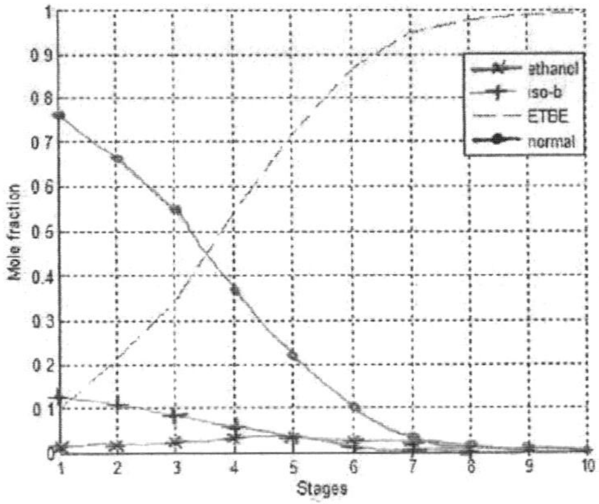

Figure 5. Liquid mole fraction profile when ETBE fed in stage 2

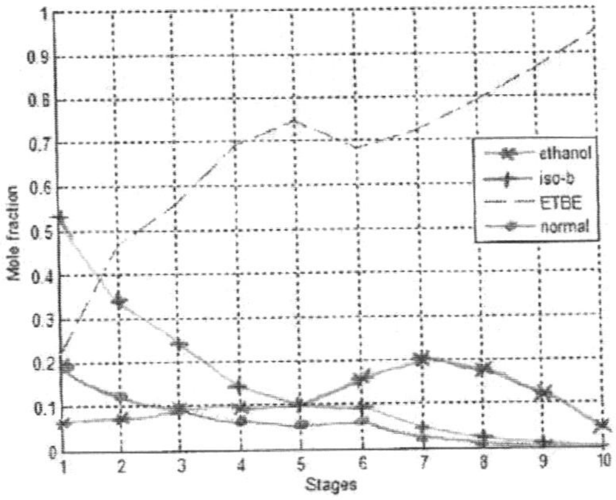

Figure 6. Liquid mole fraction for ETBE fed in stage 2 with four reactive stages

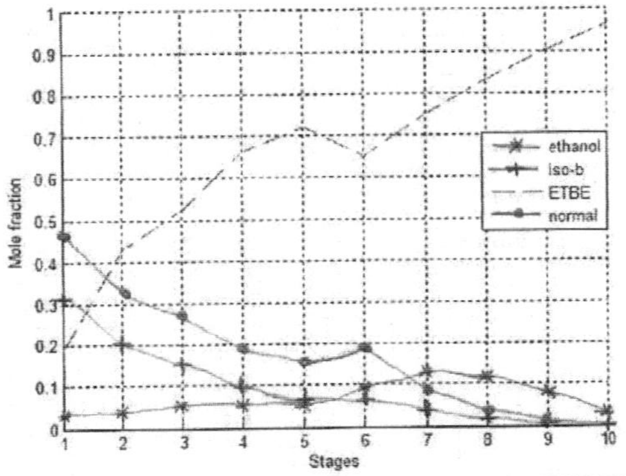

Figure 7. Mole fraction for ETBE & ETOH fed in stage 2 with four reactive stages

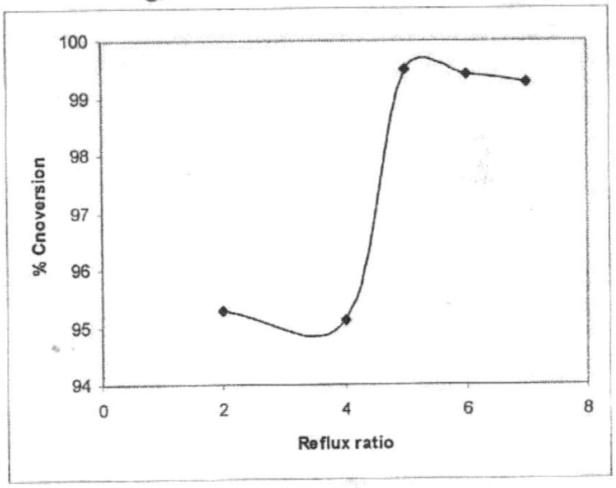

Figure 8. Effect of changing reflux ratio on conversion, 3 reactive stages with ETBE fed to stage 2

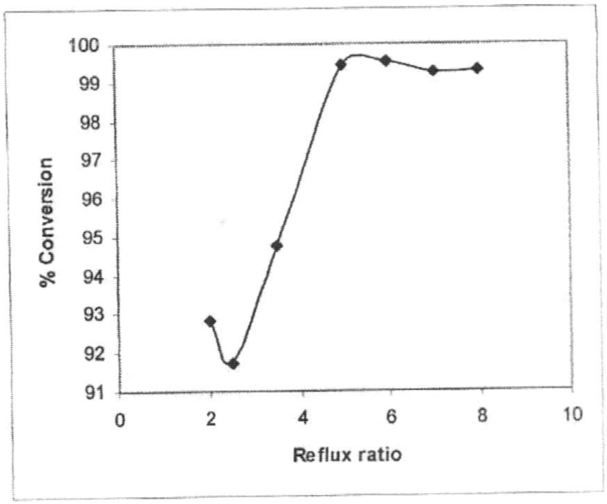

Figure 9. Effect of changing reflux ratio on conversion, 3 reactive stages with ETBE & ETOH fed to stage 2

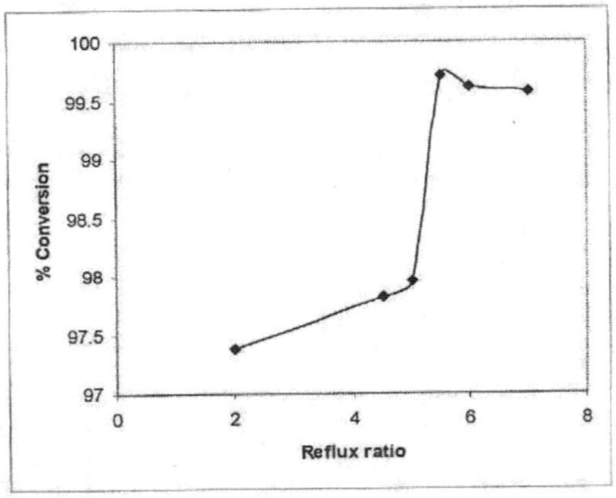

Figure 10. Effect of changing reflux ratio on conversion, 4 reactive stages with ETBE fed to stage 2

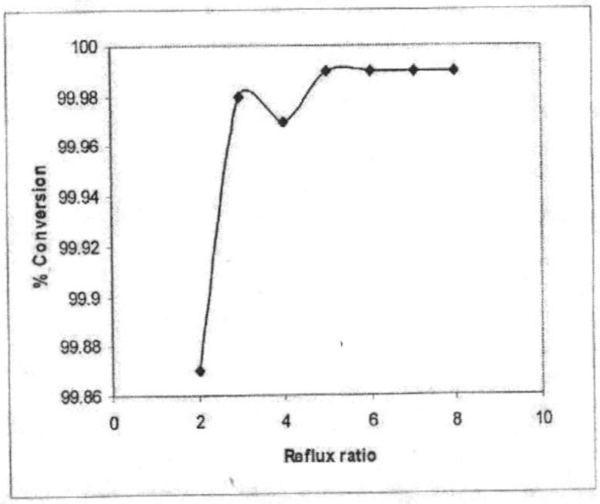

Figure 11. Effect of changing reflux ratio on conversion, 4 reactive stages with ETBE & ETOH fed to stage 2

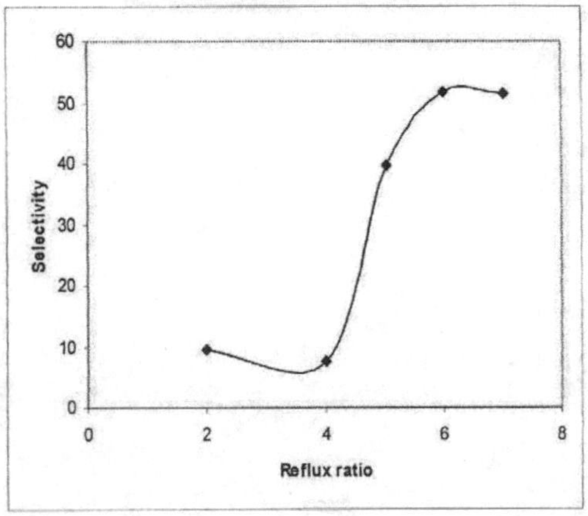

Figure 12. Effect of changing reflux ratio on selectivity, 3 reactive stages with ETBE fed to stage 2

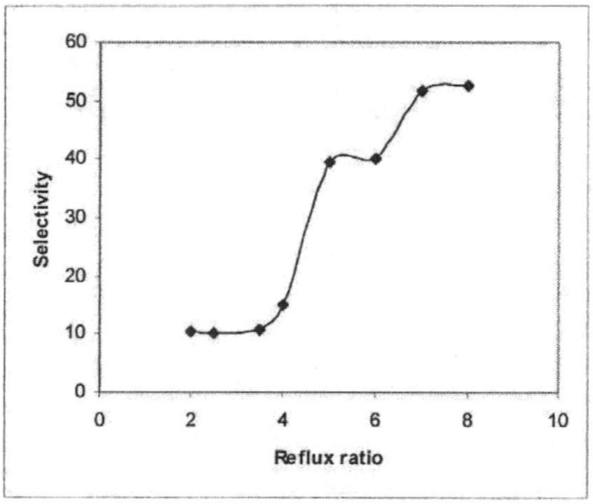

Figure 13. Effect of changing reflux ratio on selectivity, 3 reactive stages with ETBE & ETOH fed to stage 2

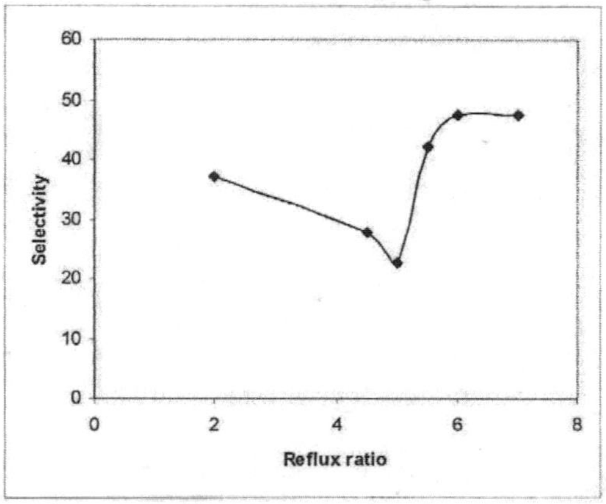

Figure 14. Effect of changing reflux ratio on selectivity, 4 reactive stages with ETBE fed to stage 2

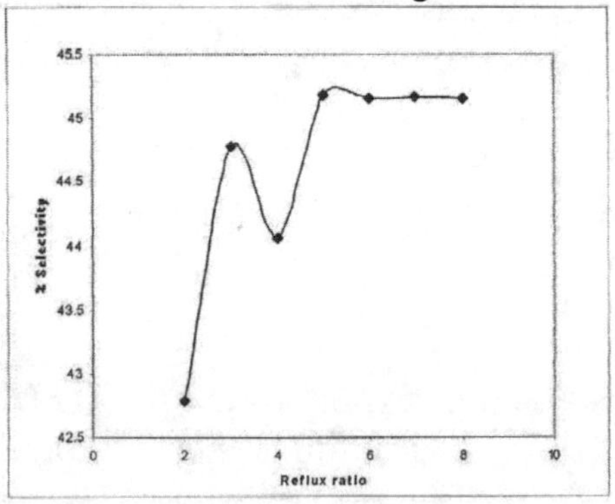

Figure 15. Effect of changing reflux ratio on selectivity, 4 reactive stages with ETBE & ETOH fed to stage 2

Study on mass-transfer performance of a new novel Higee——Rotating Jet Bed

Li Yuming, Wang ying, Ji Jianbing, Wang Guangquan

(College of Chemical Engineering and Materials Science, Zhejiang University of Technology, Hangzhou 310014, China)

Abstract: The Rotating Jet Bed(RJB) has been developed by Zhejiang University of Technology, as a more novel type of gas-liquid mass-transfer equipment than Rotating Zigzag Bed(RZB). The rotor of RJB is composed of perforated concentric baffles fixed to a rotating disc. The top of the baffles is in the grooves of the stationary disc, preventing short cut of the gas. Experiments of total reflux distillation with three types of Higee were carried out with ethanol-water systems at atmospheric pressure. The three types of Higee are RJB with rotor of mesh baffles, RJB with rotor of shutter baffles and RZB. The results showed that the number of theoretical plates per meter of RJB, lower than that of RZB, increased with the increase of rotational speed, and increased firstly and then decreased with the increase of gas superficial kinetic energy F-factor. The maximum number of theoretical plates per meter is 40.7 for RJB with rotor of mesh baffles and 40.0 for RJB with rotor of shutter baffles. The specific pressure drop of RJB was far less than that of RZB. When the gas kinetic energy F-factor was 1.1 $m.s^{-1}(kg.m^{-3})^{0.5}$, the ratios of the specific pressure drop of RJB with rotor of mesh baffles and shutter baffles to that of RZB were 10.0% and 13.6%, respectively. The gas-phase film volumetric mass-transfer coefficients of RJB with rotor of mess baffles and shutter baffles were correlated with the Reynolds number, Grashof number and Schmidt number. The calculated value agreed well with the experimental value, and the average deviation was less than 15%.

Keywords: distillation, rotating jet bed, mass-transfer, specific pressure drop

1. Introduction

The Rotating Jet Bed(RJB) has been developed by Zhejiang University of Technology, as a more novel type of gas-liquid mass-transfer equipment than Rotating Zigzag Bed(RZB)[1], as showed in Fig. 1. The rotor of the RJB is comprised of a rotating disc and a stationary disc. A series of concentric circular sheets are fixed on the rotating disc, which is called the rotational baffles. The rotational baffles are perforated, rotating with the lower(rotating) disc driven by a motor through a vertical shaft. There exists a series of concentric circular grooves on the stationary disc. The top of the rotational baffles enter into the grooves, no contact with the grooves. The perforated rotational baffles offer the flow channels for gas and liquid phase. And the grooves on the stationary disc form the labyrinth seal, preventing the gas from making a short cut. At the eye of the rotor, a special novel liquid distributor is fixed to the rotating disc, rotating with the rotating disc, which is called the rotating liquid distributor.

The liquid is fed through the rotating liquid distributor as jets or droplets onto the most inside rotational baffle. The liquid attaches and flows over the perforations of rotational baffle radially outwards

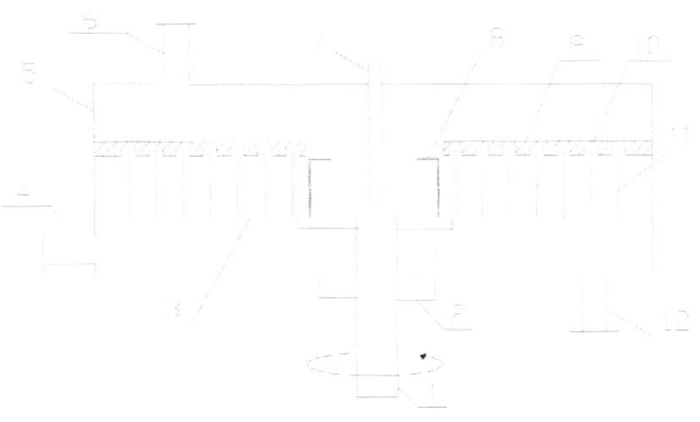

Fig.1 Schematic diagram of rotating jet bed(RJB)

1-vertical rotating shaft; 2-dynamic seal; 3-rotating disc; 4-gas inlet; 5-casing; 6-gas outlet; 7-liquid inlet; 8-rotating liquid distributor; 9-stationary disc; 10-grooves on the stationary disc forming labyrinth seal; 11-rotational baffle; 12-liquid outlet by the centrifugal force as fine jets or droplets. The liquid leaves the most outside rotational baffle, collected by the casing wall and leaving the RJB through the liquid outlet at the bottom of the casing.

The gas is tangentially introduced into the casing, flowing through the perforations of the most outside rotational baffle, and is forced to flow radially inwards countercurrent to the liquid by pressure driving force. The gas leaves the RJB through the gas outlet. The mass transfer takes place in the perforations of the rotational baffles, between two adjacent rotational baffles, and between the rotor and the casing.

The RJB has the advantages of low pressure drop, high flux, high efficiency and low power consumption, overcoming the disadvantages of excess high pressure drop and excess high power consumption of RZB.

In the paper, a total reflux distillation experiment was carried out for the study of the performance of RJB and RZB.

2. Experiment

A total reflux distillation experiment was carried out with ethanol-water system at atmospheric pressure. The performance of the RJB and RZB was studied and the structures of the RJB and RZB were illustrated in Fig.2.

The axial height of the RJB rotor is 47mm. And the RJB rotor is composed of 7 rotational baffles with their diameter of 140mm, 160mm, 185mm, 205mm, 230mm, 250mm, and 270mm. The rotational baffles of RJB employed two type of structure of mesh baffle and shutter baffle, as shown in Fig.3. The mesh baffle is perforated with diamond holes of 0.8mm in side length. The shutter baffle has plenty of shutter holes of 35mm in length, 10mm in width and 20°in opening angle.

The axial height of the RZB rotor is also 47mm. And there are 9 rotational and 10 stationary baffles for the RZB rotor. The rotational and stationary baffles are 43mm and 30mm in height, respectively. The upper part of the rotational baffles is perforated with holes of 1.5mm in diameter.

The experiment process of the total reflux distillation experiment was shown in Fig.4.

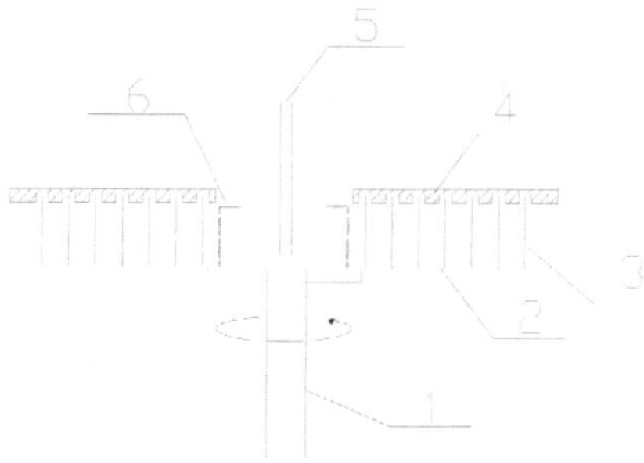

(a) rotor of RJB

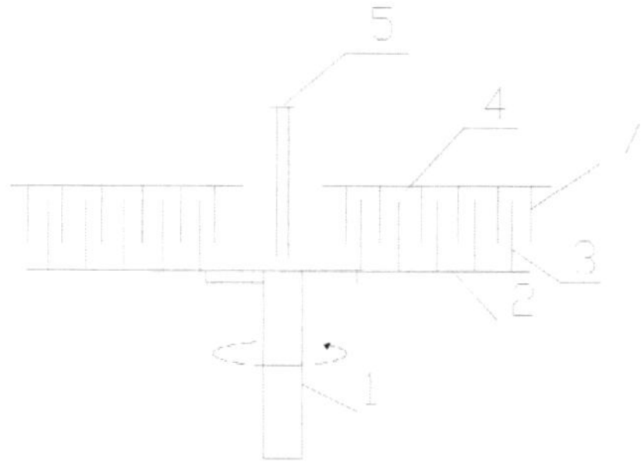

(b)rotor of RZB

Fig.2 Schematic diagram for rotor

1-rotating shaft; 2-rotating disc; 3-rotating baffles; 4-stationary disc; 5-liquid inlet; 6-liquid distributor; 7-stationary baffles

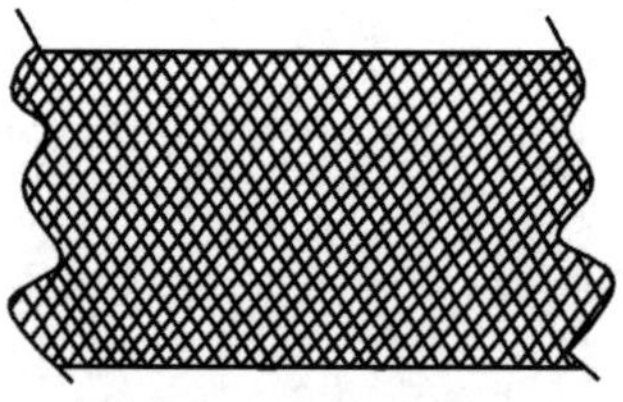

(a) mesh baffle in RJB rotor

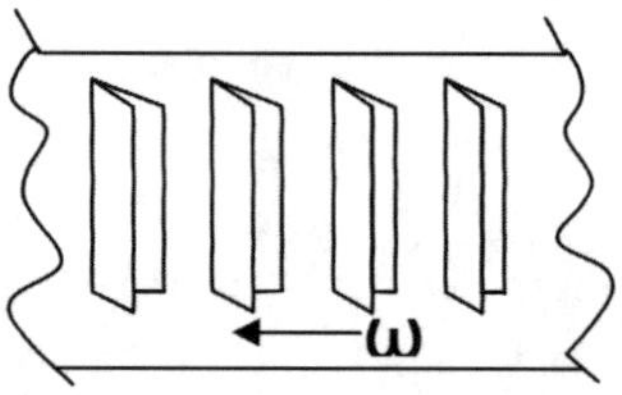

(b) shutter baffle in RJB rotor

Fig.3 Structure of baffles in RJB rotor

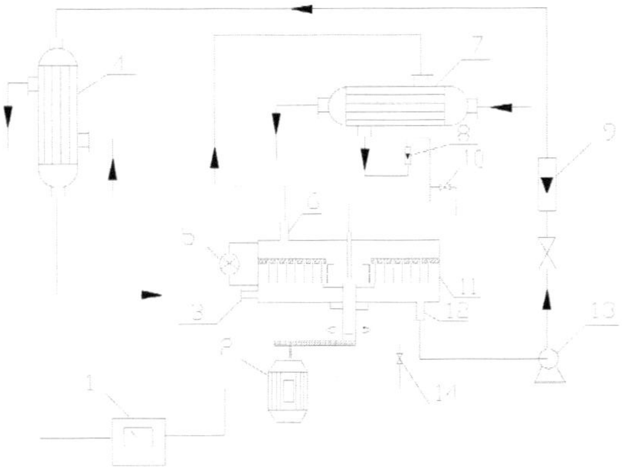

Fig.4 Experimental setup of distillation in RJB/RZB

1-frequency modulator; 2-motor; 3-gas inlet; 4-reboiler; 5-pressure difference sensor; 6-gas outlet; 7-condenser; 8,9-rotameter; 10-top sample; 11-rotating jet bed(RJB); 12-liquid outlet; 13-pump; 14-bottom sample

The liquid was vaporized into the vapor in the reboiler using heat-conducting oil. The vapor flowed into the RJB/RZB through the gas inlet. Leaving the RJB, the vapor is condensed into the liquid in the condenser. The liquid flowed into the RJB/RZB through the liquid inlet, leaving the RJB/RZB and pumped into the reboiler.

The pressure drop between gas inlet and outlet was measured with a pressure difference sensor. The reflux rates of liquid from the condenser were measured with a rotameter. The liquid samples at liquid inlet and outlet was taken to determine the liquid concentration and the number of theoretical plates.

3. Result

3.1 Effect of gas superficial kinetic energy F-factor on the number of theoretical plates per meter

The gas superficial kinetic energy F-factor of RJB/RZB is defined as $Q_G\sqrt{\rho_G}/(2\pi r_1 z)$, where Q_G is vapor rate, m^3/s; ρ_G is vapor density, kg/m^3 ; r_1 is the inner diameter of the rotor, m; and z is the height of the rotor, m.

The number of theoretical plates per meter is defined as $N_T/(r_2 - r_1)$, where N_T is the number of theoretical plates; r_2 is the outer diameter of the rotor, m.

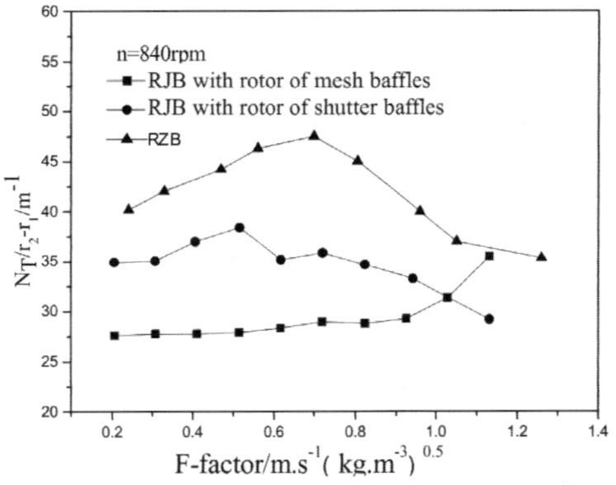

(a) rotational speed of 840rpm

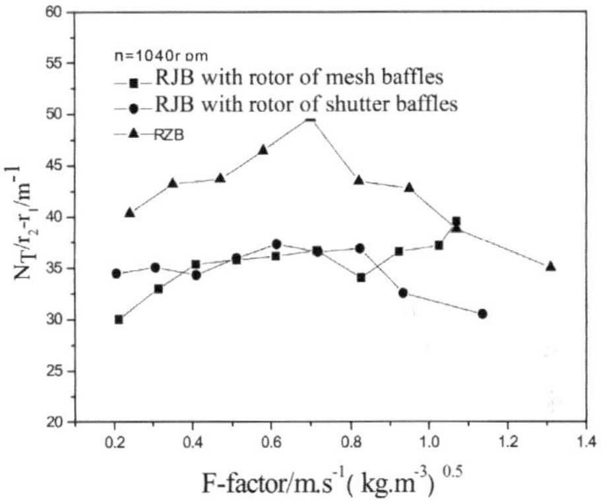

(b) rotational speed of 1040rpm

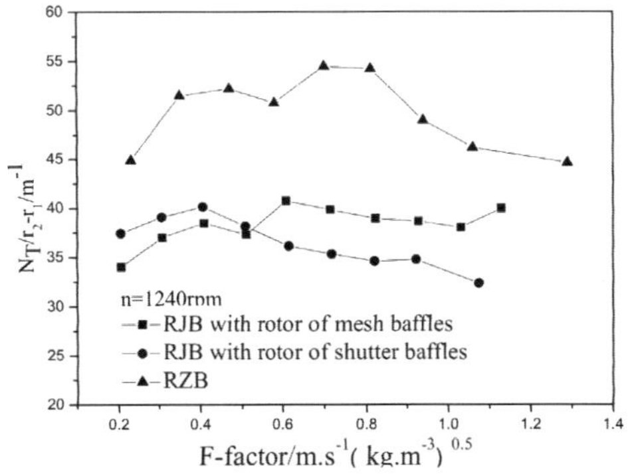

(c) rotational speed of 1240rpm

Fig.5 Effect of gas superficial kinetic energy F-factor on number of theoretical plates per meter

The effect of gas superficial kinetic energy F-factor on the number of theoretical plates per meter was shown in Fig.5.

The number of theoretical plates per meter, $N_T/(r_2-r_1)$, of RJB was lower than that of RZB, occupying 72.3~89.1% of that of RZB. However, $N_T/(r_2-r_1)$ of RJB still reached 30~40; and the maximum $N_T/(r_2-r_1)$ is 40.7 for RJB with rotor of mesh baffles and 40.0 for RJB with rotor of shutter baffles.

The number of theoretical plates per meter, $N_T/(r_2-r_1)$, of RJB increased firstly and then decreased with the increase of gas superficial kinetic energy F-factor; and $N_T/(r_2-r_1)$, of RJB increased with the increase of rotational speed.

3.2 Effect of gas superficial kinetic energy F-factor on the specific pressure

The specific pressure is defined as $\Delta P/N_T$,

where P is the pressure drop of RJB/RZB, Pa.

The effect of gas superficial kinetic energy F-factor on the specific pressure were shown in Fig.6.

The specific pressure drop, $\Delta P/N_T$, of RJB was far less than that of RZB. When the gas kinetic energy F-factor was 1.1 $m.s^{-1}(kg.m^{-3})^{0.5}$, the ratios of $\Delta P/N_T$ of RJB with rotor of mesh baffles and shutter baffles to that of RZB were only 10.0% and 13.6%, respectively.

The specific pressure drop, $\Delta P/N_T$, of RJB increased with the increase of gas superficial kinetic energy F-factor and rotational speed.

3.3 The gas-phase volumetric mass-transfer coefficients of RJB

For the packed tower, the relation of the number of theoretical plates per meter, N_T/H, and the height of a transfer unit(HTU), H_{OG}, is

$$\frac{N_T}{H}=\frac{1}{H_{OG}}\cdot\frac{\lambda-1}{\ln\lambda} \quad (1)$$

For the rotating bed, modifying the eq(1) is

$$\frac{N_T}{r_2-r_1}=\frac{1}{H_{OG}}\cdot\frac{\lambda-1}{\ln\lambda} \quad (2)$$

The height of a transfer unit(HTU), H_{OG}, multiplied by the number of transfer units (NTU), N_{OG}, is

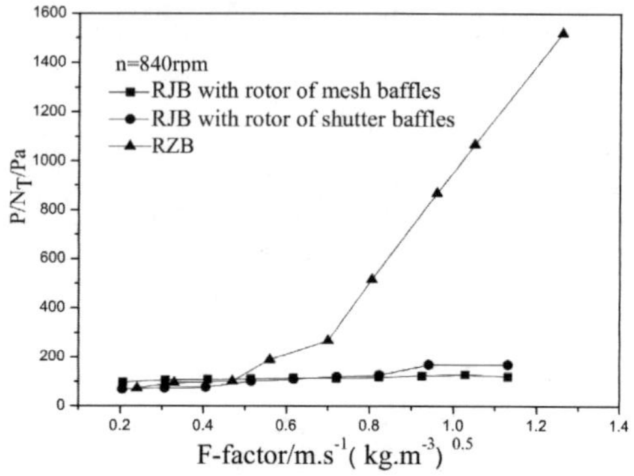

(a) rotational speed of 840rpm

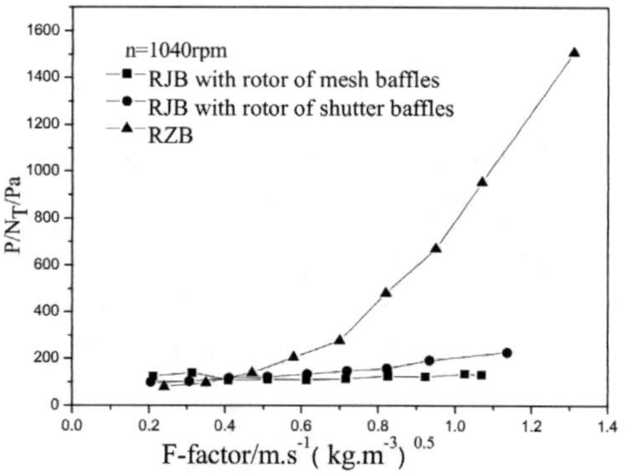

(b) rotational speed of 1040rpm

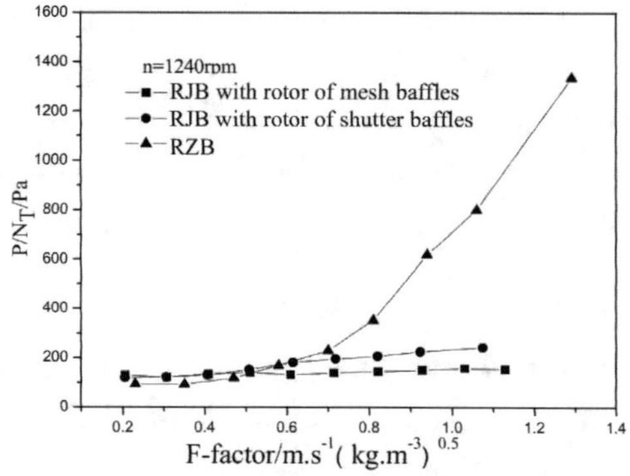

(c) rotational speed of 1240rpm

Fig.6 Effect of gas superficial kinetic energy F-factor on specific pressure drop

$$r_2 - r_1 = H_{OG} \cdot N_{OG} \qquad (3)$$

Substituting eq(2) into eq(3) is

$$N_{OG} = \frac{r_2 - r_1}{H_{OG}} = N_T \cdot \frac{\ln \lambda}{\lambda - 1} \qquad (4)$$

where $\lambda = m \cdot (V / L)$, and m is average slope of vapor-liquid equilibrium curve, V/L is the ratio of vapor to liquid.

The gas overall volumetric mass transfer coefficient is [2]

$$K_G a_e = \frac{Q_G}{z\pi\left(r_2^2 - r_1^2\right)} \cdot N_{OG} \qquad (5)$$

Substituting eq(4) into eq(5) is

$$K_G a_e = \frac{Q_G}{z\pi\left(r_2^2 - r_1^2\right)} \cdot N_T \cdot \frac{\ln \lambda}{\lambda - 1} \qquad (6)$$

From eq(6), the gas overall volumetric mass transfer coefficient may be obtained.

Due to the gas-film control in the distillation, the gas overall volumetric mass transfer coefficient, $K_G a_e$, is equal to the gas film volumetric mass transfer coefficient, $k_G a_e$. The correlation of $k_G a_e$ is [3]

$$k_G a_e = C(a_p D_G / d_p) Re^x Gr^y Sc^z \qquad (7)$$

where Reynolds number is $Re = \frac{G_G}{a_p \mu}$, Schmidt number is $Sc = \frac{\mu}{\rho D}$, Grashof number is $Gr = \frac{d_p^3 \rho^2 r \omega^2}{\mu^2}$; a_p is surface area of 216m^2/m^3 for RJB with rotor of mesh baffles and 256 m^2/m^3 for RJB with rotor of shutter baffles; d_p is hydraulic diameter of 0.77mm for RJB with rotor of mesh baffles and 15.5mm for RJB with rotor of shutter baffles; and the exponent, z, of Schmidt number is assumed to -1/3[7].

Using the experimental data, The correlations of $k_G a_e$ of RJB with rotor of mesh baffles and shutter baffles are as follows:

RJB with rotor of mesh baffles:

$$k_G a_e = 10.327(5.077) Re^{1.13} Gr^{0.30} Sc^{-1/3}$$

RJB with rotor of shutter baffles:

$$k_G a_e = 128.98(0.419) Re^{0.935} Gr^{0.053} Sc^{-1/3}$$

The calculated value agreed with the experimental value, with the greatest error of 14.05% and average error of 4.98% for RJB with rotor of mesh baffle; and with the greatest error of 9.58% and average error of 4.07% for RJB with rotor of shutter baffle.

4. Conclusion

RJB is superior to RZB. Although the number of theoretical plates per meter of RJB is slightly lower than that of RZB, the specific pressure drop of RJB was far less than that of RZB.

The number of theoretical plates per meter of RJB reached 30~40; and the specific pressure drop of RJB is only 200Pa.

The gas-phase film volumetric mass-transfer coefficients of RJB were correlated with the Reynolds number, Grashof number and Schmidt number.

References

[1] G.Q. Wang, Z.C. Xu, J.B. Ji, Performance of a rotating zigzag bed——A new HIGEE, Chemical Engineering and Processing: Process Intensification, v 47, n 12, p 2131-2139, **2008**

[2] Yu-Shao Chen, Chia-Chang Lin, Mass transfer in a rotating packed bed with viscous newtonian and non-Newtonian fluids, v 44, n 4, p 1043-1051, **2005**

[3] Kelleher T , Fair J R Distillation studies in a high-gravity contactor[J] Industry & Engineering Chemical Reseach **1996** , 35 : 4646-4655.

AUTHOR INDEX